SpringerBriefs in Water Science and Technology

SpringerBriefs in Water Science and Technology present concise summaries of cutting-edge research and practical applications. The series focuses on interdisciplinary research bridging between science, engineering applications and management aspects of water. Featuring compact volumes of 50 to 125 pages (approx. 20,000-70,000 words), the series covers a wide range of content from professional to academic such as:

- Timely reports of state-of-the art analytical techniques
- Literature reviews
- In-depth case studies
- Bridges between new research results
- Snapshots of hot and/or emerging topics

Topics covered are for example the movement, distribution and quality of freshwater; water resources; the quality and pollution of water and its influence on health; and the water industry including drinking water, wastewater, and desalination services and technologies.

Both solicited and unsolicited manuscripts are considered for publication in this series.

More information about this series at http://www.springer.com/series/11214

Chaitanya Baliram Pande

Sustainable Watershed Development

A Case Study of Semi-arid Region
in Maharashtra State of India

 Springer

Chaitanya Baliram Pande
All India Coordinated Research
Project for Dryland Agriculture
Dr. PDKV Akola
Akola, Maharashtra, India

ISSN 2194-7244   ISSN 2194-7252   (electronic)
SpringerBriefs in Water Science and Technology
ISBN 978-3-030-47243-6   ISBN 978-3-030-47244-3   (eBook)
https://doi.org/10.1007/978-3-030-47244-3

This Springer imprint is published by the registered company Springer Nature Switzerland AG
The registered company address is: Gewerbestrasse 11, 6330 Cham, Switzerland

Dedicated to my parents and all my teachers

Preface

Semi-arid regions of India are today suffering from lack of sustainable water and groundwater. There is a growing demand throughout the country for sustainable watershed development, management, and planning. The demand is more significant in the rain-fed and drought-prone area of Maharashtra, where watershed management is poorer and groundwater is limited. The area studied in this book is situated in the Akola and Buldhana districts of Maharashtra. The primary focus of the book is the improvement of sustainable watershed development and water resource and environmental management. Conservation measures involve the preparation and implementation of various projects to improve the management of watershed functions that affect the plants, animals, and human communities within the watershed boundary. Rapid, accurate, and cost-effective geospatial technologies that can be used for planning include remote sensing, GIS, and GPS.

The last eight to nine decades of professional involvement in the field of watershed management, which have involved land use, geology, hydro-geochemistry, and hydrology, and technologies including groundwater modeling, remote sensing, and GIS, have created the need for a handbook on sustainable watershed development, management, and planning for the use of hydrology, soil and water conservation and watershed professionals, as well as Ph.D. and graduate students. The book includes original field research carried out by the author for various watershed development, GIS, and remote sensing studies. All sources are acknowledged where appropriate. References appear at the end of each chapter. My efforts will be successful if this book is useful for those for whom it is written.

Akola, India   Chaitanya Baliram Pande

Acknowledgments

I am grateful to all the authors of the many research publications mentioned in the list of references for this book. The information contained in this important literature source provided the essential basis for conducting my research and writing this title. I express my gratitude to those teachers, researchers, and organizations for their contributions that reinforced my own knowledge. I am thankful to my colleagues, especially all co-authors of this book. Without their help and cooperation this research would not have been possible. Finally, I express my gratitude to my parents, sister, and brother who have been a perennial source of motivation and confidence for me. I also want to thank my wife Priti Pande for her understanding and full support while I worked on this research.

Chaitanya Baliram Pande

Contents

About the Author

Chaitanya Baliram Pande received his BCS in computer science in 2008, MSc in geoinformatics in 2011, and Ph.D. in environmental science in 2016. He has many years' experience in the field of water resources, remote sensing, land use, land cover, watershed management, hydrogeology, hydrological modeling, groundwater quality, groundwater modeling, geology, hyperspectral remote sensing, and GIS applications in natural resources management, environmental monitoring, and assessment. He has over 60 publications to his name, including three book chapters, as well as numerous conference papers and journal articles. His research interests are in the areas of integrated water resources management in river basins, land use, land cover, watershed management, hydrogeology, hydrological modeling, groundwater quality, and groundwater modeling.

Chapter 1
Introduction

Abstract In this chapter the fact that water is the most important sustainable and scarce natural resource for agriculture, manufacturing, and domestic use is discussed. The planning and management of sustainable water supplies are major time-critical challenges for any country that need to be resolved due to ever-increasing demand. Major, medium-sized, and minor irrigation and water storage projects completed over the last fifty years in India have made a significant contribution to the country's growth. It is useful to produce environmental indicators that can be combined with collateral and social indicators. In the study of sustainable watershed innovation, multi-resolution, multi-spectral techniques have been appropriated for rapid, unbiased mapping and monitoring of natural resources in both space and time. Remote sensing and collateral data enable sustainable watershed creation, prioritization, and management in erosion-prone and drought areas. Using remote sensing and GIS techniques, precise information can be provided on planning of spatial distribution, land use, soil, vegetation density, forest, geology, and water resources.

Keywords RS · GIS · Watershed · Sustainable water resources · India

1.1 Background

With a population in excess of one billion (342 million in 1947), 16% of the world's population, India is now one of the world's most heavily populated nations. By 2030 its population is expected to exceed 1.43 billion. India's agriculture sector faces many problems related to climate change, which directly affects soil and water quality, and water management. Annual precipitation in India is not sufficient due to environmental changes which also have a direct impact on groundwater levels. Growing demand is leading to further development of water resources (Arnold et al. 1990). Physical developments are contributing to economic advances by increasing comfort, convenience, and potential income for individuals (Murthy 2000).

Sixty-eight percent of India's land is used for rain-fed farming (Wani et al. 2003, 2009). Large water irrigation schemes can cover millions of hectares of farmland, while secondary water storage schemes, such as agricultural water reservoirs, and small water conservation facilities can also be used to meet the agriculture and drinking water needs of small villages. A large number of new watersheds are being planned, together with sustainable water management systems to monitor surface and groundwater supplies, together with general policies for the critical preparation, growth, and administration of these programs (Ravindran and Jayaram 1997; Khadri and Pande 2016; NWDPRA 1992; Pande et al. 2017, 2019a). The main features of watershed development are regulation of sustainable water and soil resources; among the major hydrological factors are geology, runoff, landfill, soil erosion, soil cover, topography, vegetation, and rock types (Kulkarni and Deolankar 1995; Khadri and Deshmukh 1998; Khadri and Moharir 2016).

The watershed conservation strategy and the socio-economic growth of the country are integral to natural resource management planning (Mocanu et al. 2013). Among the projects on the watershed and natural resource system initiated by government agencies and others in India, a large number concern the development of water management including such matters as groundwater sources, water quality, streams, stormwater runoff, water rights, and sustainable watershed planning. For watershed planning, various features such as stream length, geomorphometrics, ground, contour, and water belt, as well as advanced technologies such as remote sensing, GIS techniques, and satellite data are widely studied. Satellite data can be used to determine the geography of natural resources and the resulting maps used to better plan watershed growth (Khadri and Pande 2014a, b). Raster and vector data have been successfully used by policymakers, landowners, land-use authorities, conservation authorities, environmental and stormwater experts, and assessors (Pande et al. 2018a, b).

Groundwater is a secondary source of drinking and irrigating water, ensuring that soil and water are abundant and unpolluted. It is important for sustainable food production, water resources management, defensive irrigation and water supplies, agricultural crop patterns, soil erosion, drainage efficiency, agricultural land capacity, mitigation of climate change, and economic development. Groundwater is created by surface runoff and its supply depends on factors such as stream forms, density, and watershed catchments (Maggirwar 1990). Groundwater is a better option for irrigating and drinking water and a healthy source of surface water. There have been some issues with the use of soil water by animal and human users, including agriculture and industrial development, and improvements in the climate, and the inaccessibility of soils as a result of overutilization or flooding of soil waters today (Maggirwara and Umrikar 2009).

Maharashtra state, situated within the peninsular shield zone of the country, is a zone of basalt and hard rock in western central India. About 94% of its total geographical area is underscored by hard rocks and the remaining 6% contains scattered sediments and alluvial deposits. In general, hard rock is ideal for water supply. Approximately 80% of the state contains basaltic lava fluxes with excessive alluvial flux limited to nearby rivers and streams. The thin alluvial deposits and stream paths of gravel and sand have been extensively developed. A large part of Maharashtra state suffers from drought and changes to the water table, which directly impact sustainable crop yields, irrigation, and energy systems (Moharir et al. 2019). The Deccan Traps is one of the world's biggest volcanic provinces, more than 2 km deep, with an estimated area of 518,000 km^2 of continental flood basalt. The Deccan was formed 60–65 million years ago by a massive series of volcanic eruptions (Khadri et al. 1996).

Significant changes in the pattern of rainfall in any year in Maharashtra mean severe drought and lack of rainfall in certain areas, devastation and flooding in others, with the distribution of rain being unusually uneven across the state (Wani et al. 2002). The saltiness of the groundwater presents critical problems in Akola and Buldhana, where an integrated investigation and analysis is required that takes into account drinking water and rural requirements.

This book focuses on the challenges and problems of sustainable watershed growth in Maharashtra region. The primary objective of the study is to use satellite raster data to separate upstream sustainable development of water sources from current irrigation management methods and groundwater regeneration approaches.

1.2 Study Area

The study area is located in the Akola and Buldhana districts of Maharashtra, located at 76° 46'11″ E and 20° 40′ 36″ N (Fig. 1.1). The total study area is 5429 km^2

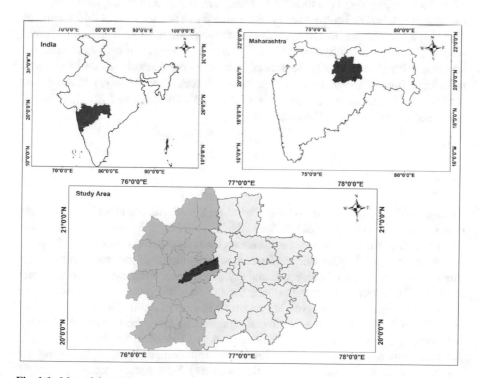

Fig. 1.1 Map of the study area

(Akola) and 9661 km^2 (Buldhana). The river basin, with an area of 328.25 km^2, is located from 240 to 580 m above mean sea level. The geology of the area consists of young and older alluvium and basalts. Seventy-five percent of the population are engaged in agriculture activities. The major crops are cotton and soybean during the kharif season, and gram and wheat in the rabi season. Most of the crops depend on seasonal rainfall. A major contributor to annual total rainfall is the southwest monsoon. Average annual rainfall of 750–850 mm occurs in the usual mountain range. The Katepurna, Morna, Man, Vidrupa, Shahanur, Van, and Nirguna are the main tributaries of the Purna River. Unconsolidated sediments, black and red soils, and the basalt rocks of the Deccan Traps cover the majority of the watershed area (Moharir et al. 2020).

The southern part of the investigation zone is covered by alluvium soil; this area is strongly affected by salt with high pH ranges in the groundwater. The Akola and Buldhana districts are bounded on the north by the steep slopes of the southern foothills of the Gavilgarh range. The watershed area lies between two major lines, the Purna lineament and the Kadam lineament. The WNW–ESE Purna lineament, marking the northern boundary of the basin, follows the Purna river and is visible for more than 200 km from the east to the west of Jalgaon, converging with the Tapi lineament from south to south Amravati. The Kadam lineament extends up to 280 km in a NW–SE direction; it marks the southern part of the basin and is named after the Kaddam River, the direction of which is determined by the fault lineament (Moharir et al. 2017). Faulting has been identified in a few sections of this lineament. The Purna, Katepurna, Morna, and Mankarna rivers all display many fertile alluvial tracts. The sloping southern portion is covered with open scrub forest, trees, and scrubland. Physiographically the area contains minor and major erosion forms, with geological landforms such as plateau, butte, dissected moderately, pediment, and pediplain. The drainage patterns flow to the south and west and are dendritic to sub-dendritic (Khadri and Pande 2014a, b).

1.2.1   Climate

The study area is situated within the central region of Maharashtra and the climate consists of four cycles: a cold season from November to February; a hot season from March to May; the southwest monsoon from June to September; and the post-monsoon season from October to November. The climate of the study area, except during the monsoon, is mostly warm and dry. The highest temperature in the summer season is 48 °C and during the winter, the lowest temperature is between 70 and 80 °C. The highest relative humidity recorded is 16% in April. The average overall humidity is 38%. During the southwest monsoon in June, the maximum recorded wind speed is approximately 14.7 km/h, and in December the lowest 4.4 km/h. Average annual wind speed is 8.6 km/h. Precipitation is around 45% during the southwest monsoon season. Precipitation records from the study area cover nearly 50 years and there are eleven rain gage stations. Annual rainfall for the districts of

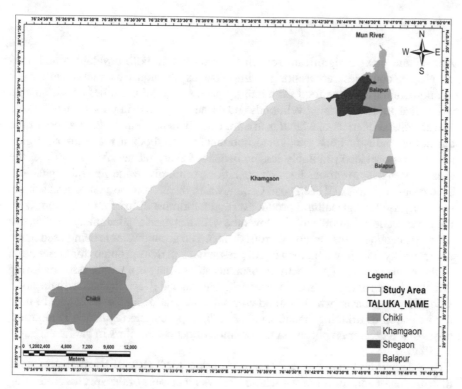

Fig. 1.2 *Taluka* boundary map

Akola and Buldhana is 750–1000 mm, lowest in the northwest and rising towards the southeast of the area (Fig. 1.2). In the west and southeast parts, in the villages Andri and Sirpur, it is observed that the chances of receiving regular rainfall are high, with both stations enduring extreme drought conditions between 1 and 4% of the year, except for Balapur. The average likelihood of extreme drought occurring in Akola is 4%. The extreme northern and northeastern parts, which experience moderate and severe conditions of drought for over 20% of the year, are known as drought zones.

1.2.2 Forest

Forests are a significant global resource supporting millions of plants and animal species. Vegetation cover is very important to prevent soil erosion. Perennial trees such as babul (*acacia arabica*) and neem (*azardirachta indica*) are seen. Few kate-savar (*bombax* species) are observed and fruit-bearing trees are also rare. Ber (*ziziphus jujuba*), *tamari dus indica*, amla, and other shrubs are found in parts of the study area.

1.2.3 Agriculture

Agriculture plays a significant role in the economy in both developing and non-developing countries, representing an important export industry for an economically strong nation as well as food for a hungry nation. Agriculture relies on sustainable watershed production, and livelihoods and economic growth in villages in the Akola and Buldhana districts of Maharashtra are entirely dependent on the success of the agriculture industry. Food yields are important for everybody in the semi-arid region and food is provided profitably for the benefit of every farmer. Almost 80% of the area is given over to various farming operations, primarily by the general population. Two crops are harvested each year. Groundwater, the most important natural source, is widely used in agricultural production and for almost 80% of all irrigation. Over the past decade, the number of shallow tube wells has risen exponentially, leading to a dramatic decline in the supply of groundwater. The groundwater is being used much more rapidly than it can be recharged, presenting a serious problem for agriculture. In the Akola and Buldhana districts, groundwater salinity is a critical issue requiring effective management in order to meet drinking water and farming needs. In the north side of the river basin, which is filled with salt water, the groundwater cannot be used for irrigation or drinking (Pande et al. 2019b). Appropriate regulation is required, as well as a greater emphasis on the creation of deeper aquifers in the first shallow aquifer region.

An agricultural map has been prepared using ARC GIS 10.1 software to show distribution of soil, storm or wasteland. It shows that most of the area for cultivating wasteland or stormy ground is situated around the periphery of the basin while the forest is dense in the far south of the basin towards the Chikali district. Crops grown in the present area of research are mainly kharifs, predominantly cotton, sorghum, jawar, pigeon pea, green gram, and black gram. Wheat, sunflower, and rabi plants typically take in residual moisture. The majority of the land is irrigated using water channels or tanks. In the southern and northeastern parts of the basin, there is maximum irrigation density for groundwater (Khadri and Moharir 2013; Khadri et al. 2013).

1.2.4 Physiography

The horizontal Deccan rock flows are physiographically roughly divided into numerous scarps and cliffs in semi-arid areas of the southern regions, and low-lying plains in the northeast (Plate 1.1). There are many erosion-like porches in the area. The horizontal arrangements of laval flows result in a reasonable degree of lithological uniformity, significantly simplified by changes induced by secondary processes, such as weathering and denudation. Differential weathering forces have been shown to wipe out thick lava piles. The area under study falls into the high focal land district with medium convergence of magma flows to the Purna river basin tributaries showing the mature phase of growth, according to the physiographic community.

Plate 1.1 A view of exposed Deccan basaltic rocks at Undri village

The area is recognized as a moderate morphogenetic region, and geomorphic analysis indicates that erosional forms are prevalent over land depositional forms. Spheroid weathering and fluvial erosion slope analysis of the main geomorphic processes in this region shows that there are flat crests of an interim slope with a very constant point of the plain.

A contour map was created using ARC GIS 10.1 to show the physiographical detail; the contour lines were defined by joining points of equal height. In the western part of the area, the landscape in the Pimpri, Pala, Nirod, and Kinhi villages up to Takarkheda is shown as a series of constructive relief features (Fig. 1.3 and Plate 1.1). The village of Nirod in the Pimpri hills is shown as on a modest slope, but the slope appears steep on the field information and satellite information.

1.3 Objectives of the Study

1. To introduce watershed and water resource management using soil and water conservation structures.
2. To prepare thematic maps for sustainable water resource development planning using remote sensing and GIS technology.
3. To develop plans for land and water resources management and watershed development.

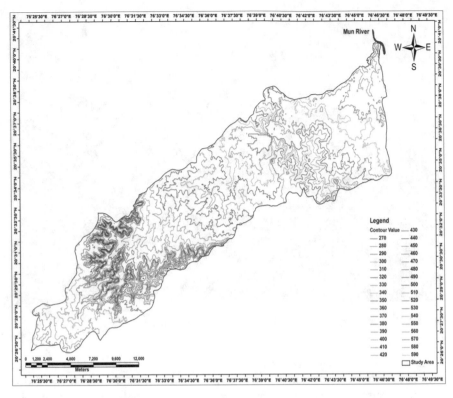

Fig. 1.3 Physiography map of Mahesh river basin

1.4 Scope of the Study

The production of water supply is an evolving cycle that needs to take account of increasing demand (Arnold et al. 1990). Conservation of storage is a significant problem in sustainable water resource development. The present research seeks to prepare for the expansion of water supplies in a specific region. The essential goal of this study is to highlight some of the major problems in the planning of water resources, combining traditional and modern methods and collecting secondary information from satellite data within the study area to generate various thematic layers to support action planning and execution.

The critical information gap associated with the salinity issue can only be addressed through comprehensive laboratory-based soil analysis and interpretation of GIS and RS satellite images followed by the development of natural resource maps of semi-arid watersheds (Rao 1991). This research seeks to raise awareness of complex salinity and sustainable watershed management initiatives taking place in the public domain, using digital methods and a multidisciplinary approach with a view to enhancing socio-economic development and watershed management (Pande and Moharir 2015; Pande et al. 2018a, b).

1.5 Remote Sensing and Collateral Data

Remote sensing and GIS technology is a computerized decision-support resource that uses a combination of spatial and non-spatial technologies for processing, storing, manipulating, and displaying data. It is an effective way of planning and handling large data groups with all the advantages of a computer environment: precision, consistency, and lack of machine error. In order to plan and create new sustainable watershed management systems, topographical, lithological, and hydrological data require to be collected, systematized, and analyzed. New effective techniques and methods, such as remote sensing, GIS, and machine learning have been developed, and these technologies may lead to accurate, rapid, and cost-effective analysis for the development and planning of sustainable watershed management (Pande et al. 2018a, b).

A broad range of efficient planning and management tools are opened up by GIS and remote sensing techniques. Remote sensing data have been combined and linked to field data to provide a single hybrid database for optimal watershed planning and management (Solanke et al. 2005). Space-borne remote sensing technology is a powerful resource that can provide repeated multi-spectrum time and space information (Lillesand et al. 2004). GIS data can detect issues and potential used in conjunction with traditional data for ridgeline demarcation, characterization, and priority needs assessment, and can identify the appropriate land and water storage sites for dam control and reservoir control (Patode et al. 2017).

The basic topographic details in this study, such as slope and drainage, were generated using ARC GIS and Survey of India (SOI) software and the toposheets are on a scale of 1: 50,000. Other software used in the development of maps included ARC GIS, ARC Details, MS Office, Google Earth and SRTM, RADAR data, SOI toposheet, and IRS-ID, LISS-3 and PAN Merged. Geocoded IRS 1D satellite imagery (LISS-III and PAN combined) was used to prepare thematic maps that are essential for sustainable watershed planning. False-color composite (FCC) for IRS 1D LISS-III and PAN merged data were used to identify soil types, soil erosion, and land use in the watershed area (Khadri and Pande 2013).

1.5.1 Georeferencing

Various thematic maps were digitized in this study using Survey of India toposheets, satellite data, and village charts. Rough georeferencing is achieved by comparing real-word coordinates with particular projection, unit and zone data in a process known as georeferencing. A map is projected for filtered georeferencing during image rectification (warping) and the specified map projection transformed with the relevant data. Land Satellite TM images were obtained from the central USGS data of our study area.

Georeferencing, a process whereby two- or three-dimensional objects are positioned in space, consists of rotating, translating, and displaying an image to suit a given size and part. The following series of actions need to be performed. Open the ARC Map by clicking on the attached data tool, selecting toposheet, and clicking 'ok.' In the georeferencing tool, right-click the grid corners of the toposheet, input the x and y coordinates, and save. Click the ARC catalog to pick and edit a new image file, then right-click on the spatial reference. Click on the coordinate geographic system, followed by the arc toolbox and then the data processing tool, the prediction, and the raster, then click on the raster and the projected raster.

1.6 Conclusion and Recommendations

This chapter has discussed the purpose, scope, and processing of remote sensing data, which is critical for the sustainable management of watersheds. The findings indicate that proper preparation of the watershed has a successful impact on the management of natural resources, aquifer areas, soils, sustainable land recharges, and access to groundwater (Khadri and Pande 2015a, b).

In view of the major financial investment, it is important that the strategic preparation and development programs for the watershed project should be successful. Recent advances in technology have greatly contributed to the inclusive, integrated resource planning and management approach. Entire streams of more readily accessible data can now be exchanged, improved information-gathering methods used, and water supply models developed to produce additional data and watershed cycle information for comprehensive, multilevel analysis. The NGOs, universities, government and other agencies, and individual scholars participating in the management, monitoring, and evaluation of project development in the semi-arid regions of Maharashtra, India, will benefit from this sustainable action plan.

References

Arnold JG, Williams JR, Nicks AD, Sammons NB (1990) SWRRB—a basin scale simulation model for soil and water resources managemen. Texas A&M Press, College Station, TX

Khadri SFR, Deshmukh MS (1998) Amravati Dist. (MS) with emphasis on water resource management, the workshop on GW recharge and management in Wardha River basin. Organized by Water Irrigation Commission, Government of Maharashtra

Khadri SFR, Moharir K (2013) Detailed morphometric analysis of Man River basin in Akola and Buldhana districts of Maharashtra, India using Cartosat-1 (DEM) data and GIS techniques. Int J Sci Eng Res 4(11)

Khadri SFR, Moharir K (2016) Characterization of aquifer parameter in basaltic hard rock region through pumping test methods: a case study of Man River basin in Akola and Buldhana districts Maharashtra India. Model Earth Syst Environ 2:33

Khadri SFR, Pande C (2013) Hydrogeological investigation of PT-6 watershed in Akola District, MS, India using Remote Sensing and GIS Techniques with reference to watershed management.

International Conference on Interdisciplinary Applications of Remote Sensing and GIS Published in International Journal of Scientific and Engineering Research, 4(12)

Khadri SFR, Pande CB (2014a) Morphometric analysis of Mahesh River basin exposed in Akola and Buldhana districts, Maharashtra, India using remote sensing & GIS techniques. Int J Golden Res Thoughts 3(11)

Khadri SFR, Pande C (2014b) Hypsometric analysis of the Mahesh River basin in Akola and Buldhana districts using remote sensing & gis technology. Int J Golden Res Thoughts 3(9). ISSN 2231-5063

Khadri SFR, Pande C (2015a) Analysis of hydro-geochemical characteristics of groundwater quality parameters in hard rocks of Mahesh River basin, Akola, and Buldhana district Maharashtra, India using geo-informatics techniques. Am J Geophys Geochem Geosyst 1(3):105–114

Khadri SFR, Pande C (2015b), Remote sensing based hydro-geomorphological mapping of Mahesh River basin, Akola, and Buldhana districts, Maharashtra, India—effects for water resource evaluation and management. Int J Geol Earth Environ Sci 5(2):178–187, May–August. ISSN: 2277-2081

Khadri SFR, Pande C (2016) Ground water flow modeling for calibrating steady state using MODFLOW software: a case study of Mahesh River basin, India. Model Earth Syst Environ 2:39. https://doi.org/10.1007/s40808-015-0049-7

Khadri SFR, Pande C, Moharir K (2013) Geomorphological investigation of WRV-1 Watershed management in Wardha district of Maharashtra India; Using remote sensing and geographic information system techniques. Int J Pure Appl Res Eng Technol 1(10)

Khadri SFR, Subbarao KKV, Walsh JN (1996) Stratigraphy from and structure of the east Pune Basalts, Western Deccan Basalt province, India. J Geol Soc Ind Mem No. 38, W. D. West Volume

Kulkarni H, Deolankar SB (1995) Hydrogeological mapping in the Deccan Basalt—an appraisal. J Geol Soc Indian 46(4):345–352

Lillesand TM, Kiefer RW, Chipman JW (2004) Remote sensing, and image interpretation. Wiley

Maggirwar CN (1990) Suitability of water harvesting, conservation, and artificial recharge techniques in relation to watershed behaviour in Maharashtra. In: All India seminar on modern techniques of rain water harvesting, water conservation and artificial recharge for drinking water, afforestation, horticulture and agriculture, Pune, Maharashtra

Maggirwar BC, Umrikar BN (2009) Possibility of artificial recharge in overdeveloped miniwatersheds: A RS-GIS approach. e-J Earth Sci India 2(II):101–110, April 2009

Mocanu M, Vacariu L, Drobot R, Muste M (2013) Information centric systems for supporting decision-making in watershed recourse development. In: 19th International conference on control systems and computer science, Bucharest, pp 611–616

Moharir K, Pande C, Patil S (2017, May) Inverse modeling of Aquifer parameters in basaltic rock with the help of pumping test method using MODFLOW software. Geosci Front 1–13

Moharir K, Pande C, Singh S, Choudhari P, Rawat K, Jeyakumar L (2019) Spatial interpolation approach-based appraisal of groundwater quality of arid regions. Aqua Jl 68(6):431–447

Moharir KN, Pande CB, Singh SK, Del Rio RA (2020) Evaluation of analytical methods to study aquifer properties with pumping test in Deccan basalt region of the Morna river basin in Akola district of Maharashtra in India, Groundwater Hydrology, Intec open Publication, UK. https://doi.org/10.5772/intechopen.84632

Murthy KSR (2000) Groundwater potential in a semi-arid region of Andhra Pradesh—a geographical information system approach. Int J Remote Sens 21(9):1867–1884

NWDPRA (1992) WARSA guidelines: national watershed development project for Rainfed areas. Department of Agriculture and Cooperation, Government of India, New Delhi, p 145

Pande CB, Moharir K (2015) GIS-based quantitative morphometric analysis and its consequences: a case study from Shanur River basin, Maharashtra India. Appl Water Sci 7(2). ISSN 2190-5487. Accessed 23 June 2015

Pande CB, Khadri SR, Moharir KN, Patode RS (2017) Assessment of groundwater potential zonation of Mahesh River basin Akola and Buldhana districts, Maharashtra, India using remote sensing and GIS techniques. Sustain Water Res Manag. https://doi.org/10.1007/s40899-017-0193-5. Accessed 8 Sept 2017

Pande CB, Moharir KN, Pande R (2018a), Assessment of Morphometric and Hypsometric study for watershed development using spatial technology—a case study of Wardha river basin in the Maharashtra, India. Int J River Basin Manag. https://doi.org/10.1080/15715124.2018.1505737

Pande, CB, Moharir KN, Khadri SFR, Patil S (2018b) Study of land use classification in the arid region using multispectral satellite images. Appl Water Sci 8(5):1–11

Pande CB, Moharir KN, Singh SK, Dzwairo B (2019a) Groundwater evaluation for drinking purposes using statistical index: study of Akola and Buldhana districts of Maharashtra, India. Environ Dev Sust (A Multidisciplinary Approach to the Theory and Practice of Sustainable Development). https://doi.org/10.1007/s10668-019-00531-0

Pande CB, Moharir KN, Singh SK, Varade AM (2019b) An integrated approach to delineate the groundwater potential zones in Devdari watershed area of Akola district, Maharashtra, Central India. Environ Dev Sustain. https://doi.org/10.1007/s10668-019-00409-1

Patode RS, Pande CB, Nagdeve MB, Moharir KN, Wankhade RM (2017) Planning of conservation measures for watershed management and development by using geospatial technology—a case study of Patur Watershed in Akola district of Maharashtra. Curr World Environ 12(3). ISSN 0973-4929

Rao UR (1991) Remote sensing for sustainable development, Photonirvachak. J Ind Soc Remote Sens 19(4):217–235

Ravindran KV, Jayaram A (1997) Groundwater prospect of Shahbad Tehsil, Baran district, Eastern Rajasthan a remote sensing approach. J Indian Soc Remote Sens 25(4):239–246

Solanke PC, Shrivastava R, Prasad J, Nagaraju MSS, Saxena RK, Baethwal AK (2005) Application of remote sensing and GIS in watershed characterization and management. J Indian Soc Remote Sens 33:239–244

Wani SP, Sahrawat KL, Sreeedevi TK, Pardhasaradhi G, Dixit S (2009) Knowledge-based entry point for enhancing community participation in integrated watershed management. In: Proceedings of the comprehensive assessment of watershed programs in India, ICRISAT, Patancheru, 25–27 July 2007. ICRISAT, Patancheru, Andhra Pradesh, India, pp 53–68

Wani SP, Pathak P, Tam HM, Ramakrishna A, Singh P, Sreedevi TK (2002) Integrated watershed management for minimizing land degradation and sustaining productivity in Asia. In Adeel Z (ed) Integrated land management in the dry areas. Proceedings of a joint UNU-CAS international workshop, Beijing, China, 8–13 September 2000. United Nations University, Tokyo, Japan, pp 207–230

Wani SP, Singh HP, Sreedevi TK et al (2003) Farmer-participatory integrated watershed management: Adarsha watershed, Kothapally India, an innovative and up-scalable approach. A case study. In Harwood RR, Kassam AH (ed) Research towards integrated natural resources management: examples of research problems, approaches and partnerships in action in the CGIAR, pp 123–147. Interim Science Council, Consultative Group on International Agricultural Research, Washington, DC

Chapter 2
Watershed Management and Development

Abstract In this chapter the watershed as a hydrogeological drainage feature that typically utilizes land and water assets is discussed. India's central government and the Maharashtra state government are introducing various water boundary improvement programs, which require a systematic methodology for sustainable watershed development and soil-related resources. Remote sensing (RS) and geographical information (GIS) systems may be used for effective control of the whole field of groundwater properties. Watershed development treatments include essential activities such as rainwater production, soil management, and water management to estimate natural protection measures. Two areas within Akola and Buldhana districts have been studied for this research. For two seasons, SOI toposheets and village maps have been completed. Supplementary information was gathered using GPS and remote sensing, including soil rates, agriculture, population, and socio-economic information. Maps including SOI toposheets and satellite images provided basic contour, drainage, soil, geomorphology, slope, and land-use information. All maps were reviewed for the design and creation of soil and water conservation assets and the sustainable development and management of water resources.

Keywords Watershed · GIS · RS · Toposheets · Natural resources

2.1 Introduction

The management of environmental resources involves various hydrology, physical-organic, socio-economic, and political dimensions. Soil and vegetation are key to the essential assets of land and water. Water is a powerful natural management tool, but also a complex one. Watersheds have critical functions affecting plants, animals, and human communities, including water supply, water quality, streams, storm runoff water, and water rights, and watershed management. Landowners, land-use organizations, conservation consultants, ecological researchers, water-use inspectors, and the public are all concerned with groundwater and surface water production and planning and the implementation of relevant projects and activities. Lectures have been

© The Author(s), under exclusive license to Springer Nature Switzerland AG 2020
C. B. Pande, *Sustainable Watershed Development*,
SpringerBriefs in Water Science and Technology,
https://doi.org/10.1007/978-3-030-47244-3_2

given covering key capability issues and the use of sustainable water technologies (Pande et al. 2019a, b; Moharir et al. 2019).

The consequences of rainfall are determined by moisture conditions and the prevailing atmosphere at the time. Therefore, due to changes in the climate as well as the spatial variability of the landmass, watershed response is non-linear. In India, the management of ecological resources in the watershed is undertaken at village level. These partnerships at the local level must be improved by management devices that are easy to understand, use integrated watershed management tools and provide information to make correct choices (Moharir et al. 2017a, b; Murthy 1998).

This chapter considers the key characteristics of a healthy watershed. It also examines the use of remote sensing and GIS technologies to build outline models for local-level planning and to establish the features of watershed sustainability. Well managed large, medium-sized, and minor irrigation projects have been an important factor in semi-arid production in India over the last 50 years. Key irrigation systems cover a large number of hectares of land, and small tanks can be used to meet the requirements of village-level communities (Khadri and Pande 2016; Khadri and Moharir 2016). Different examples have been found to account for the existence of ecological problems when such projects are applied. While an integrated watershed management approach is the best for sustainable environmental resource planning, it is rarely implemented because of its perceived lack of relevance to the production of the watershed system. Correct maintenance procedures can be overlooked (Fig. 2.1).

2.2 Integrated Watershed Management Philosophy

This chapter describes a comprehensive watershed management methodology, using a case study to assess the priority of the watershed. It explores integrated watershed management as well as the multifunctional management and planning process undertaken with a watershed investor. A team strategy is used to identify the resource challenges and concerns of the watershed, promoting and implementing the watershed development strategy with results that are ecologically, informally, and financially sustainable.

IMM has been active in India since 1970. Over the years, there have been many different implementation strategies. In 1995 the government of India, the World Bank Group, and a group of universities formally initiated sustainable, multi-spectrum watershed planning and development programs with specific targets. Following a 1999 review by the Ministry for Rural Development and the Ministry of Agriculture, a set of working approaches, priorities, plans, and regular disbursements were established in 2001, including the DPAP (Drought-Prone Area Programme), DDP (Desert Development Programme), and IWDP (Integrated Watershed Development Programme). These strategies promote the involvement of NGOs, semi-governmental organizations, private enterprises, universities, and development organizations in watersheds. These numerous projects continue to rely on groundwater production and rainwater collection methods for sustainable development (Wani et al.

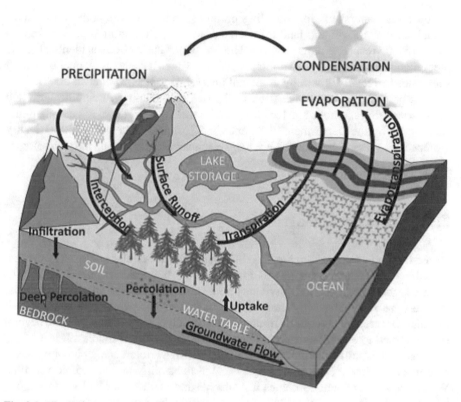

Fig. 2.1 The hydrologic cycle

2008). In order to determine the logical sequence of the proposed operations, analysis of all relevant data is necessary.

Water change is the least effective dimension in which an assessment of human effects is possible based on ecological resources. Thus, while the *Panchayat* remains the most favored part of the implementation, it is the watershed that should be used to determine results. To quantify the effects of activities carried out at the Panchayat/watershed level at a higher point in the drainage basin, a framework that maps these parts and how they are connected at Panchayat and watershed levels is required. This structure may be used by all line offices and updated by the key divisions to which ward areas are allocated in the information area. Full replication of the program will be required. This system would be feasible at local, state and national level, as well as at the river, sub-lake, and lake level (Arnold et al. 1990).

A method that meets the fundamental requirement of integrated watershed management is hydrological simulation modeling. This simulates the level in the sub-watersheds of a water and groundwater region, modeling land utilization and land cover to detect changes. The model provides a complete record of the quantity of water given to the land as a result of precipitation. The watershed is divided into separate sub-watersheds for modeling purposes. Input data are collected on individual sections of soil and land for each sub-water shaft, and these are referred to as hydrologic response units (HRUs). The process of these HRUs is identical to the precipitation inputs. The model includes all the water balance components at each sub-watershed level at monthly or annual intervals per day (Wani et al. 2003).

This present project, which used GIS-based technology to combine with and improve existing methods and technologies for village-level development planning, was funded by the Department of Science and Technology in New Delhi (UNDP). Water science researchers and NGOs were responsible for reporting the technologies presented here during the UNDP project.

2.3 Watershed Management

Watershed management is defined as the appropriate use of all ecosystem resources, including land, water, and air, to ideally establish current demands and minimize ecosystem degradation. Civil servants, environmental scientists and financial planners are all involved in watershed management. The philosophy of watershed management researches cultural, financial, and official issues both within and beyond the watershed to establish reasonable use of land and water resources. It is a multi-level method enabling resource analysis to establish the tolerance limits of potential resource exploitation, ensuring its physical sustainability and economic viability. Watershed management combines instruments and methods for soil, hydrological, biotic, and vegetative resource enhancement within the ecological environment of a stream basin to meet people's sustainable development needs. Watershed planning is a modern approach to managing soil and humidity analysis for an enhanced farm perspective similar to land use according to capacity (Murthy 2000).

The main goal of sustainable planning and growth is soil and water conservation; in this context the history of soil and water and their value must be established. A supportive structure that recognizes the role of the past within the present is needed. The decision-making process also needs to balance financial support with other operating costs, and to socialize the existing dynamics of the market and the expectations of long-term ecosystem sustainability. Sustainable manufacturing focused on health, energy, and hygiene are essential components for soil and water operations. Consequently, land and water planning and development are based on the scientific use of natural resources. The perfect geographical unit is the development of rain and land communication in the watershed area (Pande et al. 2018b).

2.4 What is a Watershed?

A watershed is a 'natural hydrologic element' draining towards specific rivers at a particular point under the earth's surface. The term 'watershed' is used synonymously with catchment area and drainage pipe. A water change region is called a separate drain. A change is a land area that can be reached at a particular point along a lake. A riverbank area is distinguished by the highest rises surrounding the river. A drop of water from the area can drain to another water bay. The watershed is also a valuable unit for economic analysis and the consideration of various physical changes related to asset use and growth. Hydrologic units are appropriate to evaluate available resources and to plan and use various development programs (Dwivedi et al. 1988).

A significant proportion of the watershed is used by people and livestock. A watershed is not defined simply by the element of water but is also a social-political object and an environmental object that plays a crucial role in defining food, social, and economic safety and in providing the human community with life-support facilities (Wani et al. 2008).

2.5 Watersheds and Stream Orders

Watersheds and stream networks are defined by the whole of the landscape. There is a specific terminology for streams (Horton 1945). First-order streams are defined as unbranched stream networks (Fig. 2.2). A second-order stream is one with two or more streams of the first order and the third-order stream has two or more streams of the second order. The water source that the drainage system uses takes on the same stream order. Although there is no evidence that streamflow and watershed characteristics are defined by stream orders, the terminology helps to position a

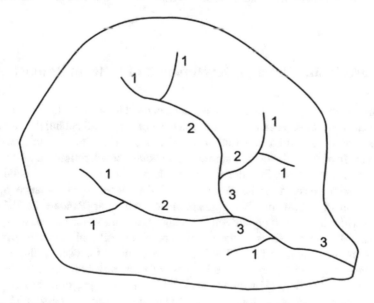

Fig. 2.2 Stream order

stream channel or a bay in a whole river basin water drainage system. Watersheds and their environment, where they are present, determine the nature and routes of water flows. In addition, watershed hierarchies in a river system typically influence water flow scales (Pande and Moharir 2018).

2.6 Watershed Assessment

The hydrological system (see Fig. 2.1) responds to climate change and land-use change. It is essential to examine watershed boundaries and to evaluate several watershed metrics, including human activity. Physical parameters need to be established, such as the basin length, stream length, water field, river slope, slope, water drainage order, and drainage density. More detailed and accurate data for the creation of the vegetative cover, geology, and soil map are obtained using GIS and remote sensing data (Pande et al. 2019a, b). Hydrological unit codes (HUCs), based on work in the United States, are used to identify watersheds. The four stages of hydrology are identified, beginning with 21 main geographical areas. This includes one of the major river basins in a particular region with another river basin chain. The key areas are divided into 221 subregions of 378 components; the final number of components is 2264. HUCs are used in thematic mapping to define watersheds or other features of the land and related characteristics such as climate, vegetation, geology, soils, land use, and topography. The hydrology properties of HUCs can be predicted based on various characteristics of similar HUCs. For the specific applications covered in this book, these techniques and methods of assessing watershed characteristics are required.

2.7 Sustainable Use and Development of Natural Resources

There are several ways in which climate changes need to be better and more sustainably managed to conserve natural assets and meet existing and potential water needs. The use of assets is not based strictly on the physical and natural characteristics of waterways. Institutional, socioeconomic, and social considerations should be fully integrated into structures in order to achieve human, economic, and social goals, such as the social base of rural communities and the concept of governance. Specific examples can show how these variables relate to each other (Pande et al. 2017).

The world's increasing population is making land and water increasingly scarce resources.. Human reactions to these deficiencies are difficult to change and can have real environmental effects. The increasing demand for water induced by the expansion of populations and by the increase of monetary growth will continue to pose prominent questions (Vorosmarty et al. 2000).The changing climate and climatic conditions create uncertainty for the management of land and water assets. The extent to which freshwater sources will change with environmental and climate change, and the location of such changes, is uncertain.

2.7.1 Land Scarcity

In many developing countries, land scarcity has been exacerbated by rural poverty. Forests are cleared to grow crops, steep highland areas are cultivated, and fragile fields overgrazed to satisfy the need for food and natural resources. Watersheds also cause erosion and further rising land productivity which in turn leads to wider and more intensive land use. In other situations, unacceptable irrigation practices to improve agricultural productivity have resulted in a reduction of land productivity due to salinization. People in both uplands and downstream areas are affected by the decline of wetlands. The depletion of agricultural land production suggests that land usage of 8.7 billion ha worldwide is unsustainable. Almost 25% of the soil, forests, forests, and grasslands have been degraded since mid-1900, with 3.5% of the 8.7 billion ha worldwide, almost 25% being degraded (Khadri and Pande 2014a, b).

2.7.2 Water Scarcity

One of the biggest environmental challenge facing the world in the twenty-first century is recognized as water scarcity. In March 2001, when the United Nations celebrated the World Day for Water, water speakers argued that demand for sustainable groundwater with surface water was 15–20% higher than the world's available surface water and that two-thirds of the world's population will experience serious water shortages in the next 25 years. Even the 9000–14,000 km^3 of freshwater worldwide required in the near future to allow people to develop, will be insufficient in the world's semi-arid region (Rosegrant et al. 2002; Khadri et al. 2013a, b).

2.8 Remote Sensing and GIS for Integrated Watershed Management

Remote sensing (RS) and GIS technologies are a collection of non-spatial and spatial data-mingling techniques and decision-support systems that can analyze, manipulate, and display geographical data. They represent methods for designing and managing large-scale spatially wide data frames, with all the benefits of the CPU environment: precision, consistency, and lack of computational errors. The long-term objective of RS and GIS applications to watersheds is to understand a novel development strategy (Khadri and Pande 2014a, b). Initial analysis is carried out based on data collected and categorized on slope, soil erosion, surface topography, and aquifers, and hydro-meteorological information (Moharir et al. 2017a, b). The latest tools and technologies such as groundwater flow modeling, aquifer mapping, land-use modifying software, and hydrological modeling with satellite data represent a rapid, cost-effective analysis that can be used to produce plans relating to sustainable watershed growth and management (Fig. 2.3).

Innovations in space-borne remote sensing have created tools to provide convincing spatial, multi-spectral planning data. GIS information can be used in connection with standard range description, classification, priority assessment, identifiable facts, potential, and management assessment. For example, remote sensing and GIS has been used by specialists in groundwater management activities for rain-water harvest (Anbazhagan and Ramasamy 2006). Remote sensing data combined with field analysis knowledge provides an ideal one-of-a-kind hybrid database for watershed managers (Solanke et al. 2005). The successful use of spatial remote sensing information properly combined with other financial insurance information under GIS conditions will indicate clear local sustainable development solutions.

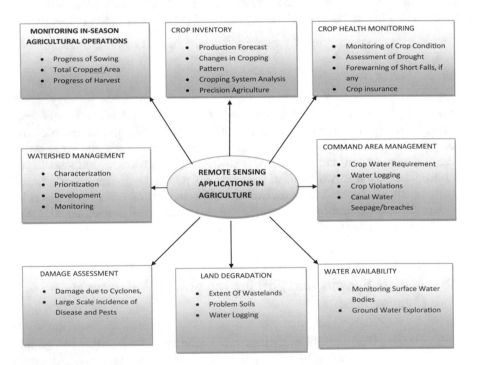

Fig. 2.3 Watershed management systems

2.9   Types of Watershed

Watersheds are categorized according to scale, presence, shape, drainage density, geology, slopes, drainage type, and land-use pattern (Pande 2014).

a.   Macro watershed (>50,000 ha)
b.   Sub-watershed (10,000–50,000 ha)
c.   Milli-watershed (1000–10,000 ha)
d.   Micro-watershed (100–1000 ha)
e.   Mini-watershed (1–100 ha).

2.10   Mini-watershed Concept

A mini-watershed is a homogeneous geo-hydrological unit covering 50–2500 ha or 5–25 m^2. A mini-watershed can include up to four villages. Surface water, ground-water, and water conservation techniques should be planned in the mini-water unit to improve consumption of water-irrigated crops with *gram sabha* approval. Only then can the watershed improvement plan be presented and implemented.

2.11   Objectives of Sustainable Watershed Management

The objectives and priorities of sustainable watershed creation and planning programs are:

a.   To control and preserve soil and water from harmful runoff and degradation.
b.   The beneficial use and management of runoff water.
c.   To preserve, maintain, and enhance watershed land to increase production efficiency and sustainability.
d.   To protect and enhance water resources from the basin.
e.   To control soil erosion and reduce the effects of sediment.
f.   To rehabilitate deteriorating land.
g.   To moderate flood peaks in downstream areas.
h.   To increase rainwater infiltration.
i.   To improve and increase the production of timber, feedstock, and wildlife resources.
j.   To enhance groundwater recharge, wherever applicable.

2.12   Factors Affecting Watershed Management

Size and shape, topography, soils, and relief are among the factors affecting planning and management of the watershed. Precipitation, and rainfall rate and volume are the climatic characteristics. Land-use patterns include type and density of vegetation. Issues of social status and weak watershed management will also play an important role (Khadri and Moharir 2013; Khadri et al. 2013a, b).

2.13   Watershed Management Practices

Science plays a vital role in the effective conservation and management of accessible water supplies. Prominent watershed practices include infiltration expansion, water-holding volume extension, prevention of soil erosion, and recharging strategies. Control measures include strip cropping, pasture cropping, and the use of grasslands and woodlands for agriculture. Engineering solutions such as contour bunding, terracing, earth embankment, check dams, farm ponds, gully-monitoring structures, and rock dams, complement the effect of soil conservation measures and vegetation cover on erosion, surface runoff, and nutrient losses. Watershed characteristics that organizations seek to manipulate are water quality, drainage, stormwater flow, and water rights. Systematic methods that require a responsive mix of conventional land measurement and remote sensing technologies pave the way for primary water supply projects to be planned and run (Moharir et al. 2017a, b). Human involvement in watershed management builds an autonomous, effective system that is indispensable for sustainability. In this semi-arid region, reliable spatial and non-spatial information is vital for study of the characteristics of a watershed intended for sustainable ecological supply (Pande et al. 2018a, b).

Watershed management seeks to halt deterioration in the links between the watershed's natural resources. Exceptional use of resources in a previous area undermines the sustainability of watershed connections. Management of groundwater bodies in the micro-watershed, with their associated water supply potential, has not been systematically evaluated. Rural planning historically concerned dams, reservoirs for irrigation, and domestic control of land runoff requiring minor irrigation tanks. The majority of the tanks, however, are historical and their storage capacity has decreased. Cultivation in the control area, on the other hand, has become more time-consuming and multiple cropping has been introduced. There are two restorative approaches to preserve correct current land-use practices. The surface storage (percolation tanks) system is being installed in some places to replenish an aquifer which is under stress. Longer storage of groundwater increases the possibility of filtration. It is expected that water storage leads to percolation, filling the aquifers. Check dams typically built to protect soil conservation can be used as mini or micro percolation tanks depending on their storage capacity. A water catchment tank has a region that is opened and also a control area. The impact of a percolation tank can also correspond to the irrigation tank's control area. The absence of a sluice in a percolation tank is a major difference between the two structures. Water is no longer drawn for irrigation immediately, but may percolate the surface strata and thereby increase groundwater (Khadri et al. 2013a, b; Khadri and Pande 2015).

2.14 Capacity Building

Capacity building to boost resources for village populations, and specifically for farmers, plays an important role in the development of watershed areas in semi-arid regions, linking all stakeholders from farmers to service managers. Capacity building is a program that is not mandatory or customer oriented and that seeks to improve people's abilities to achieve their own goals on a sustainable basis (Wani et al. 2008). Lack of awareness and information on goals, strategies, and behavior of the parties involved affect the catchment's overall performance. Stakeholders are aware of the importance of numerous and varied activities. Capacity-building software focuses on the development of cost-effective rainwater harvesting approaches, conservation, the creation and use of bio-fertilizers and bio-pesticides, income-generation measures, action-based livestock, land reclamation, and key stakeholders' bazaar association.

2.15 Watershed Management Approaches

2.15.1 Integrated Approach

This approach applies combined sciences to the natural limitations of a stream location in order to achieve top-quality expansion of land, water, and plant resources to sustainably meet the basic requirements of people and animals. It seeks to enhance people's living standards by growing farmers' production capability to reach the desired income. Combined watershed planning recommends the following rainwater and soil preservation practices for the development of a sustainable watershed: soil and water harvesting activities, such as farm ponds; the artificial renewal of groundwater to increase water sources; crop diversification; use of an extended range of seeds; built-in nutrient management; and built-in pest management practices (Pande et al. 2015).

2.15.2 Group Approach

A group strategy emphasizes collective activity and participation by principal stakeholders, government and non-government organizations, and different institutions. Watershed planning should therefore involve multidisciplinary abilities and competencies. Such a system can drive remarkable change by providing timely advice to farmers, increasing their awareness and their ability to seek advice from professionals when troubles arise. Multidisciplinary skills are required in the areas of engineering, soil, water, remote sensing, GIS, geology, hydrology, botany, horticulture, animal husbandry, entomology, social science, economics, and marketing. While it is not always feasible to gather all the required aid and skill sets within a single organization, the group method provides the advantage of specialist know-how to supplement the activities of many watershed initiatives and interventions (Singh et al. 2008).

2.16 The Importance of Land-Use Planning in Watershed Development

The unequal distribution of various natural assets and the interdependence of social, animal, and human life, call for effective preparation in the production, management, and use of semi-arid land resources. Adinarayana (2008) developed a watershed management system in which agroecological characteristics were determined using the most relevant data, assessing soil erosion, and managing conservation elements (Patode et al. 2017). Data from several sources such as the NBSSLUP, remote sensing, irrigation, agriculture, forestry, rural enhancement departments, and markets were combined with geographical data structures (GIS), simulation models, and bio-econometric models to improve water policy and rainfall planning. This concept is appropriate to a variety of watersheds (Pande et al. 2018a; Wani et al. 2008).

2.17 Conclusion and Recommendation

A participatory policy for the sustainable use of natural resources to improve agricultural productivity and reduce rural poverty proposes the preparation of a small catchment or watershed. This approach, a sustainable alternative to a pure hydrduological facility, is a collaborative land and water planning method helping to enhance livelihoods in synchronization with the local environment. Sustainable and scientific use of soil-based and water-related production taking account of the interdependence of people and animals must be codified in compliance with national and scientific standards to give priority to developing-country watersheds. Over the past 30 years, India's watershed system has established a central, integrated, community-based approach to rural livelihoods through sustainable management of natural resources. The pinnacle of this strategy consists of soil and water conservation. Integrated watershed management (IWM) programs enable soil, water, and biodiversity conservation, improved production, increased family incomes, development of community resources, and the potential resilience to cope with changes including those related to local weather and globalization. The IWM system will become a growth engine for the sustainable development of dryland tropical areas.

References

Adinarayana J (2008) A systems-approach model for conservation planning of a hilly watershed centre. Studies in Resources Engineering, Indian Institute of Technology, Bombay Powai, Mumbai, India

Anbazhagan S, Ramasamy SM (2006) Evaluation of areas for artificial groundwater recharge in Ayyar basin, Tamil Nadu, India through statistical terrain analysis. J Geol Soc India 67:59–68

Arnold RW, Szabolcs I, Targulian VO (eds) (1990) Global soil change. Report of an IIASA-ISSS-UNEP Task Force on the role of soil in global change. International Institute for Applied Systems Analysis, Laxenburg

Dwivedi RS, Reddy PR, Sreenivas K, Ravishankar G (1988) The utility of IRS data for land degradation mapping. In: Proceedings, national seminar, IIRS, mission and its application potential, Hyderabad

Horton RE (1945) Erosional development of streams and their drainage basins: hydrophysical approach to Quantitative morphology. Bull Geol Soc Am 56:275–370

Jeyakumar L (2019) Spatial interpolation approach-based appraisal of groundwater quality of arid regions. Aqua J 68(6):431–447

Khadri SFR, Moharir K (2013) Detailed morphometric analysis of Man River basin in Akola and Buldhana districts of Maharashtra, India using Cartosat-1 (DEM) data and GIS techniques. Int J Sci Eng Res 4(11)

Khadri SF, Moharir K (2016) Characterization of aquifer parameter in basaltic hard rock region through pumping test methods: a case study of Man River basin in Akola and Buldhana districts Maharashtra India. Model Earth Syst Environ 2(33)

Khadri SFR, Pande CB (2014a) Morphometric analysis of Mahesh River basin exposed in Akola and Buldhana districts, Maharashtra, India using remote sensing & GIS techniques. Int J Golden Res Thoughts 3(11). ISSN 2231-5063

Khadri SFR, Pande C (2014b) Hypsometric analysis of the Mahesh River basin in Akola and Buldhana districts using remote sensing & GIS technology. Int J Golden Res Thoughts 3(9). ISSN 2231-5063

Khadri SFR, Pande C (2015) Ground Water Quality Mapping FOR Mahesh River Basin in Akola and Buldhana Districts of (MS) India Using Interpolation Methods.International Journal on Recent and Innovation Trends in Computing and Communication 3(2):113–117

Khadri SFR, Pande C (2016) Ground water flow modeling for calibrating steady state using MODFLOW software—a case study of Mahesh River basin, India. Model Earth Syst Environ 2(1). ISSN 2363-6203

Khadri SFR, Pande C, Moharir K (2013a) Geomorphological investigation of WRV-1 Watershed management in Wardha district of Maharashtra India; using remote sensing and geographic information system techniques. Int J Pure Appl Res Eng Technol 1(10)

Khadri SFR, Pande C, Moharir K (2013b) Groundwater quality mapping of PTU-1 Watershed in Akola district of Maharashtra India using geographic information system techniques. Int J Sci Eng Res 4(9)

Khadri SFR, Pande C, Moharir K (2014) Groundwater Recharge Zone Mapping. In: PTR-2 sub-watershed, Akola district, India using remote sensing and GIS techniques. Fourth international conference on hydrology and watershed management at Jawaharlal Nehru Technological University Hyderabad, Telangana state, India, Allied Publishers, ISBN-978-81-8424-952-1, 2014, ICHWAM-2014 Volume 1

Moharir K, Pande C, Patil S (2017a) Inverse modeling of Aquifer parameters in basaltic rock with the help of pumping test method using MODFLOW software. Geosci Front 1–13, May 2017

Moharir K, Pande C, Varade AM, Pande R (2017b) Morphometric analysis in Koldari watershed of Buldhana district (MS), India using Geo-informatics techniques. J Geomat 11(1)

Moharir K, Pande C, Singh S, Choudhari P, Rawat K, Jeyakumar L (2019) Spatial interpolation approach-based appraisal of groundwater quality of arid regions. In: Aqua Journal (IWA Publication) 68(6):431–447

Murthy RS (1998) Rural psychiatry in developing countries. Psychiatric Serv 49(7):9679

Murthy KSR (2000) Groundwater potential in a semiarid region of Andhra Pradesh: a geographical information system approach. Int J Remote Sens 21(9):1867–1884

Pande C (2014) Change detection in Land use / Land cover in Akola Taluka using remote sensing and GIS technique. Int J Res (IJR) 1(8)

Pande, CB, Moharir K (2015) GIS-based quantitative morphometric analysis and its consequences: a case study from Shanur River basin, Maharashtra India. Appl Water Sci 7(2). ISSN 2190-5487. Accessed 23 June 2015

Pande CB, Moharir K (2018) Spatial analysis of groundwater quality mapping in hard rock area in the Akola and Buldhana districts of Maharashtra, India. Appl Water Sci Springer J 8(4)1–17

Pande CB, Khadri SFR, Moharir KN, Patode RS (2017) Assessment of groundwater potential zonation of Mahesh River basin Akola and Buldhana districts, Maharashtra, India using remote sensing and GIS techniques. Sustain Water Resour Manag. ISSN 2363-5037. https://doi.org/10.1007/s40899-017-0193-5, Published online 8 September 2017

Pande CB, Moharir KN, Pande R (2018a) Assessment of morphometric and hypsometric study for watershed development using spatial technology—a case study of Wardha river basin in the Maharashtra, India. Int J River Basin Manag. https://doi.org/10.1080/15715124.2018.1505737

Pande CB, Moharir KN, Khadri SFR, Patil S (2018b) Study of land use classification in the arid region using multispectral satellite images. Appl Water Sci 8(5): 1–11 ISSN 2190-5487

Pande, CB, Moharir KN, Singh SK, Dzwairo B (2019a) Groundwater evaluation for drinking purposes using statistical index: study of Akola and Buldhana districts of Maharashtra, India. In: Environment, Development and Sustainability (A Multidisciplinary Approach to the Theory and Practice of Sustainable Development). https://doi.org/10.1007/s10668-019-00531-0

Pande CB, Moharir KN, Singh SK, Varade AM (2019b) An integrated approach to delineate the groundwater potential zones in Devdari watershed area of Akola district, Maharashtra, Central India. Environ Dev Sustain. https://doi.org/10.1007/s10668-019-00409-1

Patode RS, Pande CB, Nagdeve MB, Moharir KN, Wankhade RM (2017) Planning of conservation measures for watershed management and development by using geospatial technology—a case study of patur watershed in Akola district of Maharashtra. Curr World Environ 12(3):706–714

Rosegrant M, Ximing C, Cline S et al (2002) The role of rainfed agriculture in the future of global food production. EPTD Discussion Paper No. 90. Environment and Production Technology Division, IFPRI, Washington, DC, USA

Singh, O, Sarangi A, Sharma MC (2008) Hypsometric Integral Estimation Methods. Indian J. Soil Cons

Solanke PC, Shrivastava R, Prasad J, Nagaraju MSS, Saxena RK, Baethwal AK (2005) Application of remote sensing and GIS in watershed characterization and management. J Indian Soc Remote Sens 33:239–244

Vorosmarty CJ, Green P, Salisbury J, Lammers RB (2000) Global water resources: vulnerability from climate change and population growth. Sci J 289, 14 July 2000

Wani SP, Singh HP, Sreedevi TK et al (2003) Farmer-participatory integrated watershed management: Adarsha watershed, Kothapally India, An innovative and up-scalable approach. A case study. In: Harwood RR, Kassam AH (eds) Research towards integrated natural resources management: examples of research problems, approaches and partnerships in action in the CGIAR. Interim Science Council, Consultative Group on International Agricultural Research, Washington, DC, pp 123–147

Wani SP, Joshi PK, Raju KV et al (2008) Community watershed as a growth engine for development of dryland areas. A comprehensive assessment of watershed programs in India. Global Theme on Agroecosystems Report No. 47. International Crops Research Institute for the Semi-Arid Tropics, Patancheru, Andhra Pradesh, India

Chapter 3
Thematic Mapping for Watershed Development

Abstract This chapter exclusive thematic maps are prepared for a semi-arid zone in the districts of Akola and Buldhana, while remote sensing and GIS technologies are used in earth surface mapping and watershed growth planning. In the study of land use and cover, soil density, soil erosion, and soil drainage maps, both satellite images and field data are used. Water precipitation is evaluated in the groundwater priority area in order to prepare a sustainable water supply in the watershed.

Keywords Thematic maps · Watershed · Remote sensing and GIS

3.1 Introduction

Thematic mapping relates to a specific subject matter. A thematic map shows a particular subject of interest, such as environmental and human characteristics like population density and health concerns, in contrast to an ordinary map, which reflects a range of geological and geographical events. It also shows spatial variants and minimal geographical diversity. Object maps are used for three specific purposes. First, to provide comprehensive information on various locations. Second, to present conventional data on spatial models. Third, to identify patterns on two or more maps. On thematic maps, features such as streams or roads are represented differently. They are reference points to enhance understanding of the desired characteristics of the sector (Khadri and Chaitanya 2015a, b, c).

Remote sensing and GIS are the most efficient methods for measuring more than a few earth sources. A large number of asset maps can be created by a completely remote satellite sensing method and a composite map with diverse information can also be analyzed with the aid of GIS software. Appropriate management policies and planning must be developed and implemented in line with field requirements (Patel and Dhiraj 2019).

In this chapter, an action plan for an effective water and soil conservation site is prepared that incorporates several thematic layer systems, including soil erosion, soil depth, land efficiency, slope, land use, and land drainage. The land resource action plan has been created using zoning maps and weighting for the different thematic layers. Forestry conservation and protection systems in the action plan include communal plantations with continuous trenches, forest protection and management, reforestation, canal management, agricultural dryland with ponds, horticulture planting, intensive agriculture, irrigation, and pasturing.

A water resource development action plan for the conservation of water and soil in the watershed uses different thematic layers, such as slope, land use and property, irrigation, soil textures, ground depth, soil erosion, soil and land capacities, and weightage. The use of a zoning system is proposed as a tool for sustainable management of watersheds for water conservation and to reduce soil erosion (Patode et al. 2017a,b). Suggested water conservation structures include control dam, percolation tank, earthen nala bund, form pond, graded bund, sunken pond, roof water harvesting, and loose bouldering.

3.2 Methodology

Environmental zoning is a natural system of sustainable watershed management designed to reduce soil erosion and environmental degradation of water and soil. A watershed is a region (or area) with a well-defined water outlet and topographical boundary. It is a natural area from which water concentrated in a certain location, such as a river or reservoir, is drained. The water bottom includes a complex of soils, landforms, vegetation, and land use within a topographical boundary or water divide. The terms water bath, tank, and tub are also used analogously.

Watersheds have long been known as ideal units for development planning and implementation. Their management requires the simultaneous consideration of hydrological, pedological, and biological resources, as well as the cumulative impacts of human activities and the ecological, economic, and esthetic integrity of the many strategies and drainage systems used to mitigate them. Watershed evaluation requires a methodology that can address challenging issues, but that is easily implemented, versatile yet consistent, can be implemented on various spatial levels, and can be converted into easily articulated management explanations and decisions. For resource planning, watershed strategies require timely and reliable spatial and statistical data and it is important that analytical methods and strategies that tackle spatial and temporal variation be properly utilized.

The use of satellite remote sensing has been significantly beneficial for water management with regard to both conservation and control. Remote sensory data also allow mapping of surface water resources and systems, enabling various hydrological processes and thus water equilibrium to be researched with a reasonable degree of accuracy. In this context, ARC GIS is extremely promising for handling spatial and temporal information, and can serve as an integrative management planning tool. GIS can construct and store spatial mapping and is capable of performing multiple scenario analyses/evaluations such as simulations of physical, chemical, and biological processes supporting watershed applications. Remote space-borne multispectral sensors (e.g. LISS in IRS, and others) supply spatial and temporal data at

24-day intervals at different defined and spatial resolutions (23.5 m for MSS data, and 5.8 m for panchromatic data). This helps understand the shifts in the dynamics linked to land and water supplies. Gross is an open-source GRASS-based system (geographic resources analysis support system) which works on a Linux platform. Gross can perform spatial raster and vector analyses as well as helping to interpret and model decisions. A user-friendly user Gross interface with all the GIS and image-processing capabilities has been developed that helps decision makers and planners collect, store, process, and display spatial and temporal information, imagine spatial and temporal decisions, and coordinate and prioritize them. Gross was used to classify possible sites for growth by contrasting the real watershed management websites with digital elevation models (DEMs). DEMs reflect the continuous variation in relief across the region and have widespread application in hydrological modeling, contributing to determining steep slope, slope length, the direction of flow, watershed boundaries, and outlets.

In Anantapur district, Andhra Pradesh, integral remote sensing technology provides micro-level preparation at the village level for long-term, remote sensing techniques to tackle drought. Various drought management measures related to plant water collection systems, fodder, timber, and permanent tree cover growth were recommended, as well as soil restoration and moisture conservation measures. In hot, arid regions in Karnataka the population has been swelling, but comprehensive planning approaches have been absent. A study of urbanization in river wetlands in Bangalore city shows that a range of dams, parks, playgrounds, coach stands and solid waste dumping grounds in and around the city have been converted to residential and commercial use. The study found that urbanization of 60% of the water area has more than doubled in thirteen years at the expense of farmland/open/scrub.

Various technologies have been used to construct databases designed for developing sustainable watershed management in the semi-arid zone. In the ARC Map 10.1 system, a drainage chart is produced from satellite data with SOI Toposheet. ERDAS Imagine software creates land-use/land maps with supervised classification techniques. ARC Map 10.1 software uses the available reference facts (Fig. 3.1 and Table 3.1) to create other thematic layers such as ground, geomorphology, and slopes.

3.2.1 Land-Use and Land-Cover Mapping

Land use and land cover are critical aspects of the value of the world's land, both currently and historically, and of how its use can be safely changed. Land cover is a fundamental parameter that assesses the surface material of the earth as a significant component affecting the ecosystem's situation and function. The study of land cover helps understand the interplay between biodiversity and ecosystems. Evaluation currently plays a significant role in the field of ecological science and assists land-use management. Remote sensing information has been shown to be very useful in mapping changing land-use trends for environmental protection purposes. These

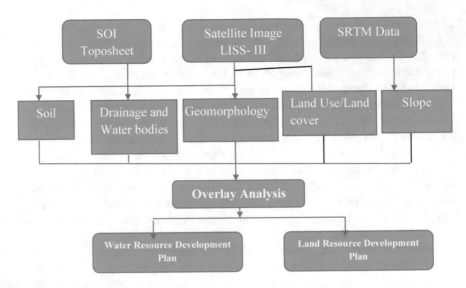

Fig. 3.1 Methodology flow chart

Table 3.1 GIS data collection and sources

S. N.	Satellite data product	Data sources
1	Satellite Data National Remote Sensing, LISS-III of IRS IC and IRS—P6 (Raw data) LISS IV Agency Government of India	Bhuvan Portal freely available satellite data in National Remote Sensing IRS ID-PAN, LISS-III of IRS1D Agency Government of India site
2	1:50,000 Toposheet, district resources map, geological formation map	Government of India, Survey of India at 1:25,000 scale
3	Maps showing existing information on the semi-arid area Akola and Buldhana district study area	M.H. State Remote Sensing Applications Centre M.H. GSDA Amravati Divisions M.H. Survey of India, Nagpur M.H. Irrigation Department, Nagpur GSDA, Amravati
4	Field data	Intensive fieldwork

changes can be identified using GIS methods even if exceptional scales/resolution are present in the resulting spatial datasets (Sarma et al. 2001).

Land is the most important natural resource in the entire ecosystem. Land use is the use of land resources by humans and associated fauna. Land cover consists of natural plants, bodies of water, rock/stone, and synthetic coverings. Changes in land cover are often the most significant environmental impacts of intense human activity. Land-use and land-cover mapping is important in the watershed landscape. Changes in land usage are not necessarily the only differences in land-use and land-cover charts, which also include in-depth and management adjustments (Pande et al.

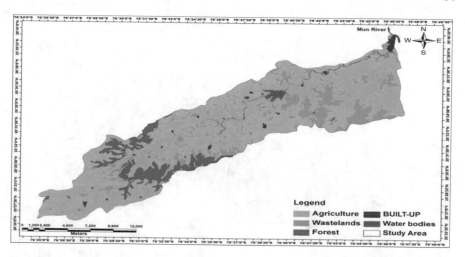

Fig. 3.2 Land-use map (Level 1)

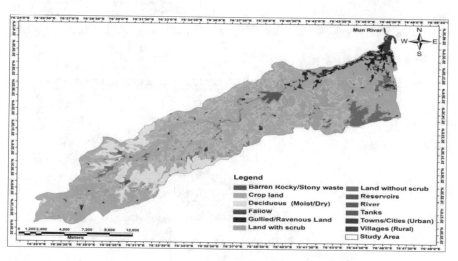

Fig. 3.3 Land-use map (Level 2)

2018; Reddy et al. 2017). They also reveal the biophysical state of the Earth and its immediate surface, including soil material, plants, and water in the watershed.

Changes to land use/cover also include modifications in the ordinary climate, either direct or indirect, which have an impact on the local ecosystem. Land utilization/alternative reporting has become a central element of contemporary strategies for natural resource management and environmental change monitoring. A land use/cover sample of a region provides data on natural and socioeconomic factors, and human subsistence and growth. Like other resources, land resources are limited by

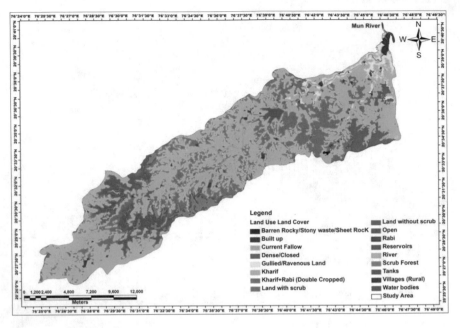

Fig. 3.4 Land-use map (Level 3)

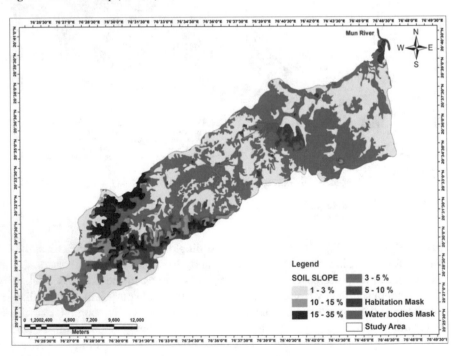

Fig. 3.5 Soil slope map

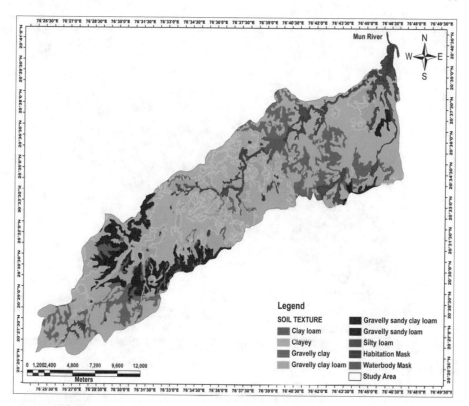

Fig. 3.6 Soil textures map

the high demand for agricultural products and the rising day-to-day stress of population. Land-use statistics and the possibilities of optimal use are therefore crucial to selecting and enforcing policy, and for planning to meet increased human needs and welfare (Moharir and Pande 2014a, b).

The study of full data on the spatial distribution of land-use groups is a prerequisite for the management and usage of semi-arid land sources. Any terrain's land-use pattern reflects the diverse physical solutions occurring on the earth's surface that influence the distribution of soils, vegetation and water prevalence in the atmosphere. In addition to environmental conditions, it is important to have timely and accurate geomorphological, geological and topographical records to develop and maintain watershed areas (Khadri et al. 2013; Kokate et al. 2014).

Satellite data has been used to establish land-use mapping in the study area (Pande and Moharir 2014; Pande and Kanak 2014). Most land is under cultivation, together with a preserved forest located in the eastern portion of the basin, and the wasteland in its south-western part. Land use and land cover captured by the satellite image are visually represented and translated into vector formats for polygon maps. Mapping groups may vary from sheet to sheet depending on the ground conditions. Analysis includes reference to relevant details, enabling compatibility with existing maps

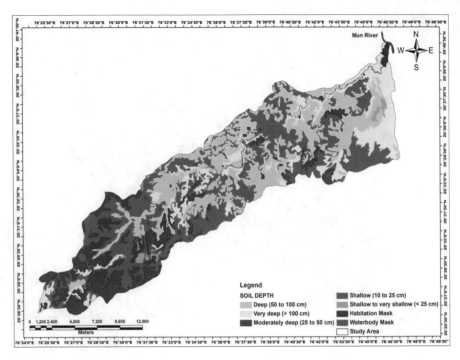

Fig. 3.7 Soil depth map

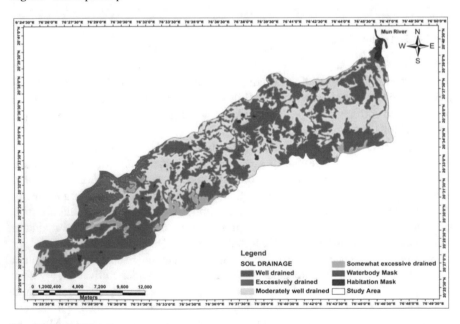

Fig. 3.8 Soil drainage map

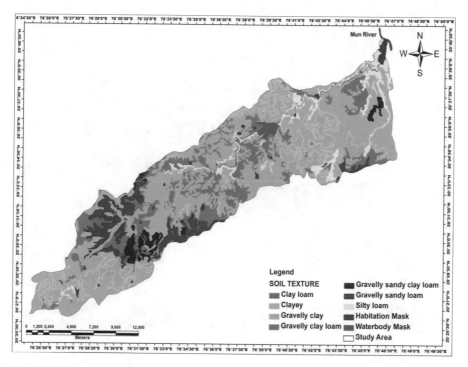

Fig. 3.9 Soil map

and knowledge of points on the Survey of India (SOI) maps, for example forest boundaries.

Three land-use and land-cover maps based on distance-sensing data are described in this analysis. The field details and classifications used are shown in Table 3.2. The level 2 and 3 maps cover kharif and rabi. All surface bodies of water (reservoirs, lakes, and tanks) are represented on an SOI toposheet survey and new constructions are identified by the current satellite data which depicts water depth. Field visits have enabled collection and interpretation of data on the ground, as well as estimates of classification accuracy. Interpretation continues until it meets the data accuracy requirements. To remove digitization errors (Pande et al. 2018b), the ARC GIS 10.1 version shapefile has been developed and updated to polyconic projection and metered co-ordinate units. The transformation was geo-referenced based on the input type file and corresponding toposheets using ground control points (GCPs) (Plate 3.3).

The use of groundwater in most watersheds is constrained by the pace of natural regeneration. Many environmental issues will emerge from unmanaged use. Soil particles are dislodged and transported by winter, water, and soil erosion, affected by factors that include climate, topography, soil use and characteristics, and vegetative cover. In many areas natural floodwater is used to counteract soil erosion and sedimentation. Wind erosion is a significant cause for concern in arid and coastal areas

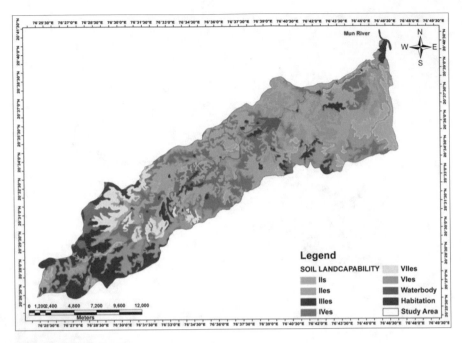

Fig. 3.10 Soil capability map

Plate 3.1 Cement nala bund at Atali village

Plate 3.2 Loose boulder structure at Undri village

Table 3.2 Classification scheme for land-use and land-cover mapping

S. N.	Level 1	Level 2	Level 3
1	Built-up	Built-up (urban) Built-up (rural)	Core urban, peri-urban Village, mixed settlement
		Mining/industrial	Hamlets and dispersed households, mining/industrial
2	Agricultural land	Transportation, cropland, agriculture plantation	Transportation, cropland, agriculture plantation
3	Forest	Forest, forest plantation	Forest, Forest plantation
4	Wasteland	Waterlogged, scrub land	Waterlogged, dense scrubland
			Open scrubland
		Sandy areas, barren rocky	Sandy areas, barren rocky
5	Water bodies	River, stream, drain	River, stream, drain
		Canal, lake, pond	Canal, lake, pond
		Reservoir, tank	Reservoir, tank

Plate 3.3   Farm Pond at Lokhanda village in Buldhana District

worldwide (especially in the tropics). Annual erosion through forestry operations and road building amounts to xxxxxx and xxxx, respectively.

Soil loss is a topographical factor for rainfall erosivity in a particular area. It is calculated using a combination of the angle and length of the slope, and is expressed in tons. l is compared with a 72.6 ft long slope, while c is compared with a slope of 9%) The crop management factor is expressed as a ratio of soil loss in the area of interest to soil loss from continuously tilled fallow under erosion control (a ratio of soil loss using erosion control practices to soil loss with farming up and down the slope). Rill and soil erosion measurements for agricultural watersheds in upland areas taken upstream of rivers or reservoirs do not include erosion from the banks of waterways and eroded sediments accumulated at the base of the slope and at other reduced flow sites. Distance formation in points and lines helps test the efficiency of hard rock wells obtain demand filters into a route chart or flow direction chart). Digitized contour interpolation and groundwater surface filtering are used to construct 2D flow network from remotely sensed data (Pande et al. 2019b).

Simple GIS models are used to measure peak runoff for environmental impact assessments. The adequacy of the drainage systems proposed by highway developers must be measured for a large number of catchments.

3.3 Precipitation and Interception

The amount, timing, and spatial distribution of water added to a watershed from atmospheric precipitation are largely beyond human control. The type, extent, and condition of vegetative land cover influence the deposition of water and the amount reaching the soil surface through interception. Dense coniferous forests and multi-storied tropical forest canopies intercept and store significant quantities of precipitation and a substantial quantity (around 30%) is returned to the atmosphere as evaporation. Moisture evaporation from soil, plant surfaces, and water bodies together with water transpired through plant leaves is called evapotranspiration. Larger-canopied plants transpire larger amounts than bare soil or smaller plants. Rainfall infiltration entering the soil through the surface, filling depressions in the soil or through rivers and streams makes up as much as 80% of the tropical watershed each year. The physical condition of the soil (porosity and hydraulic conductivity) affect the moisture content. Vegetation encourages higher infiltration than bare soils. Canopy cover reduces rainfall impact. The time and quantity of stream flow generated in a watershed is determined by characteristics such as, shape, size, channel and slope of the watershed, topography, and drainage density. The presence of wetlands and reservoirs affects infiltration and runoff. Water yield from a catchment usually increases, particularly when forests are cleared or thinned, and also when vegetation changes from deep-rooted species with high interception capacities to shallow-rooted plant species with low interception capacities.

Groundwater is the water that is accumulated beneath the soil surface. In saturated zones groundwater is very important to maintain wetness of the watershed but it is seldom found where it is most needed. Groundwater is often used as a source of fresh water and is important to sustain or revive vegetation. Groundwater that seeps into streams provides their base flow. However, in addition to the effects of fecal pollution by animals and humans, contamination with nitrates or chlorides has been reported in groundwater management studies, where pools have been dug to demonstrate the effects of nutrient contamination from agriculture.

3.4 Agriculture

Cultivation in the basin is widespread but the greatest value is concentrated in the lowlands where there are suitable water sources and other edaphic conditions. The key crops in this group are kharif, rabi, two crops, more than two cultivars, and Zaid crops. Crops were found on moderately sloping sites in the plateau areas or flat uplands (Fig. 3.2 and Table 3.3).

Table 3.3 Land-use/land-cover mapping area

Land-use classes	Area (km^2)	Percentage
Agriculture	181.90	55.41
Wasteland	109.24	33.26
Forest	30.53	9.30
Built-up land	3.0	0.91
Water bodies	6.68	1.30
Total area	328.25	100

3.4.1 Agricultural Land (Kharif Season)

The main kharif crops corresponding with the district's monsoon season (June to September) are soya, cotton, nachni, sunflower, safflower, *tur*, mung and *udid* (Fig. 3.2). The October 2014 satellite data shows an area of 166.02 km^2 under cultivation with this group.

3.4.2 Agricultural Land (Rabi Season)

Areas with reliable irrigation (surface and groundwater) are in this category, which covers an area of 1.05 km^2. The main rabi crops are wheat, gram, rabi, jowar, oilseed, and turmeric (Fig. 3.3).

3.4.3 Agricultural Land (Double Crop)

The area for the cultivation of double crops is very small, and measures approximately 13.72 km^2 (Fig. 3.4).

3.4.4 Agricultural Land (Currently Fallow)

Satellite data and field surveys find some fallow farmland, in particular in the dry mountains and plateau areas in both kharif and rabi seasons. The absence of cultivation is largely due to lack of irrigation or insufficient soil moisture. A total area of 1.09 km^2 is included as fallow in the current research (Fig. 3.4).

3.4.5 Agricultural Land (Plantations, Horticulture)

Citrus and other orchard production has a total surface area of 51.68 km^2 (Fig. 3.4).

3.4.6 Forest

Hills, upland valleys and sheltered spurs are identified in the satellite images as forest land. In Akola and Buldhana districts, only arid feeding forests are visualized due to the destructive soil conditions, hard rocky terrain, and climate factors (Fig. 3.2).

3.5 Soil Mapping

The surface layer of the earth is the product of the continuous interaction of parent matter, the surrounding atmosphere, plant and animal species, and soil elevation. It is a vital part of our ecosystem, acting as a plant anchorage and a nutrient supply. Soil is, therefore, the critical raw material for the production and planning of relevant watersheds. The study area has three types of soil. The black cotton soil lies primarily in the basin's rivers and along the riverbanks. The disintegrated volcanic basaltic lava is usually rich in clay and impregnable. It is heavy and has poor permeability. It is very dense in the lowlands and thin in the highlands. The red soil formed by the region's abundant iron compounds surrounds the black soil. In general, red soil is less fertile but is suitable for rice, millet, and fruit. In high altitudes, laterite soils are mainly found with precipitation along the ridgeline. Laterite soil is a key feature of the pebble crust.

3.5.1 Satellite and Ancillary Data

Satellite information obtained in the summer season when there is minimum crop or vegetation cover is ideal for identifying soil patterns in a semi-arid area. However, information obtained in the monsoon season combined with summer records provides better soil information under certain terrain conditions.

Additional data needed for the soil map are topographical maps, small-scale soil maps, geological maps, and climate statistics (rainfall, temperature, etc.). Ground data collection is required for the preparation of basic topographical maps (Pande et al. 2018a).

Table 3.4 Slope classes

Slope class	Slope (%)
Level to nearly level	0–1
Very gently sloping	1–3
Gently sloping	3–8
Moderately sloping	8–15
Moderately steeply sloping	15–30
Steeply sloping	30–50
Very steeply sloping	>50

3.5.2 Watershed Soil Characteristics

Soils from the soil map were used to digitize the details of the area in the semi-arid region being studied for sustainable watershed development planning. In general, satellite data identifies seven kinds of soil: asphalt, clay, stone, gravel, sand, sandy clay, and gravely clay.

The erosion of the topsoil diminishes soil fertility and causes crops to fail. As soil spectral reflectance with excessive repeatability can be quickly and easily collected, many samples can be studied to establish hydrological trends within the region. The spectrally defined indicators were adjusted to represent the pixels obtained from satellite data to spatially extrapolate the soil-physical situation in the watershed. Once established, a conditional addiction model was used to eliminate errors caused by connections between factors in the calibration model. A soil fitness index with reflective values from band 3, band 5, and band 7 was used (mean level 5%).

3.5.3 Soil Slope

Slope is an important aspect of land use. The total water runoff and soil erosion due to water flow have a marked influence on the field path. The pitch is normally expressed as a percentage. A 10% slope has a vertical fall of 10 m/100 m horizontal distance. On a 1–2% slope, soil management is increasingly required to avoid erosion problems. Slope classes are shown in Table 3.4 and mapped in Fig. 3.5 (NBSSLUP 1995).

3.5.4 Soil Texture Analysis

Soil is thick, clayey and sticky in the critical portion of the semi-arid region. The soil is a specific source of plant nutrients with higher cation levels. The earth has low hydraulic conductivity, a high degree of retrenchment, the swelling of the handling reaches 1–4 cm, with large cracks of up to 60 cm deep. The ground in the highlands is relatively saline and sodium free. In the absence of an appropriate descending gradient, the soils on the basin side are subject to soil and floor water salinization. Surface soils are sandy loam with a composition that is approximately 65% satisfactory in grain. At depths of 1 m up to 5.2 m the soil is silty clay. Soil thickness varies from place to place. In general, three types can be defined in the study area: slightly coarse in the higher levels in the south; medium in the dark black soil in the river valley; and maximum thickness in the northern part of Balapur. The deepest soil is located in the south-eastern part of the basin; in the middle and marginal or peripheral parts of the basin, shallow soil predominates (Fig. 3.6).

3.5.5 Methodology

The soil depth map was prepared using ARC GIS 10.1 and satellite data with ground reality for the extra-deep zones. Geographical software with remote sensing was used with reference to the toposheets of the SOI map as well as current geological, geomorphological, and land maps. Arc GIS version 10.1 has been developed and edited for digitalization errors. The functions are marked according to defined codes/symbols. The original file form was changed to WGS in 1984 and the coordinate device was metered. The process includes geometrical correction through the GCs found in the file and the corresponding SOI map.

Physiographical units in the form of sample strips were defined and further stratified by soil based on geological variants, landforms, parent material, elevations, slopes, aspect, vegetation, and so on. The true profiles depended on the terrain variability. Knowledge of mineralogical classifications is useful, as well as soil tempera ture, moisture regimes, and weather information. A specific local interpretation key is established through analysis of the physiographical unit/image and soil classes.

3.5.6 Soil Depth

The depth of the soil is of crucial importance for cultivation. It defines the root area and the volume of soil from which the plant can meet its requirements for water and nutrients. In some cases, depth varies across short distances (Fig. 3.7 and Table 3.5).

3.5.7 Soil Drainage

Soil drainage refers to the capacity of the soil to absorb excess water in its macropores through gravity. Soil drainage is analyzed by topography, texture, and tilt. Drainage is key to calculating crop yields in the semi-arid region and is an essential element

Table 3.5 Soil depth classes

Soil Depth (cm)m)	Series
<10	Extremely shallow
10–25	Very shallow
25–50	Shallow
50–75	Moderately shallow
75–100	Moderately deep
100–150	Deep
150+	Very deep

of the soil survey. Internal drainage, which occurs most naturally and frequently, is mainly related to texture and porosity, while external drainage relates to pooling on flat ground discovered in soils is drainage internals. The drainage conditions of a region are expressed in the soil color, low chrome mottling, and the extent of its effect in the soil. The natural drainage class refers to the extent and the moisture levels of conditions close to the soil. Soil water management through drainage or irrigation is not appropriate unless it substantially changes the morphology of the soil. The soil drainage classification (NBSSLUP 1995) is as follows (Khadri and Pande 2014).

3.5.7.1 Excessively Drained

Water is easily extracted. The soil is typically coarse-textured and highly conductive hydraulically or very shallow (Fig. 3.8).

3.5.7.2 Somewhat Excessively Drained

Water is easily extracted from the ground. The occurrence of freshwater is always very rare or very slight. In the wetlands, the soils are frequently ground textured and the hydraulic conductivity is overly saturated.

3.5.7.3 Well-Drained

Water is easily removed from the land, but not quickly. At some point, in most wetlands, water is accessible to plants in the growing season. The soils are typically free of wetness-related deep to redoximorphic aspects (Fig. 3.8).

3.5.7.4 Moderately Well-Drained

For part of the year water is removed gradually from the soil. In certain situations, hydraulic conductivity within the top 1 m layer is relatively low or lower, and occasionally rainfall is excessive or accessible (Fig. 3.8).

3.5.7.5 Imperfect

The soil is fed at a shallow depth during the growing season for long periods. Wetness greatly restricts the crop yield except for the use of artificial flux. Soils are usually one or more of the following: low or very low hydraulic conductance saturated, high water tables, additional water from drainage, or almost non-stop rainfall.

3.5.7.6 Poorly Drained

It is extracted so gradually that the soil is wet for long periods at low depths even during the growing season. The water is generally in or near the ground for long enough during the growing season to prevent growth of mesophytic plants. However, the soil is not continuously wet at plowing depth. Free water is usually located at a shallow depth.

3.5.7.7 Very Poorly Drained

Water stays close to the ground for most of the growing season. Many mesophytic crops cannot be cultivated without artificially drainage. Usually, the soils are stadium or depressed (Fig. 3.8).

3.5.8 Soil Erosion

The formation and transfer of particles of soil is called soil erosion. Soil erosion affects not only soil formation and its reputation for fertility but also land use. Agriculture and the environment are affected by soil erosion through water, wind, and tillage. Sometimes, soil erosion remains at roughly the same rate as when soil is formed. Accelerated soil erosion occurs when the rate of soil erosion is increasing and is caused by dangerous human activities, such as excessive cultivation. Soil can also be isolated or transported (including over long distances) during erosive rainfalls or windstorms.

It is not always easy to establish detailed facts about soil erosion. Research is carried out by geomorphologists, agricultural engineers, soil scientists, hydrologists, and others for the benefit of politicians, growers, environmentalists, and various individuals (NBSSLUP 1995). The following classification applies to soil erosion.

3.5.8.1   Slightly Eroded

Erosion has sufficiently altered the soil to require only minor management adjustments compared to soil without erosion; future usage and management are essentially the same. The majority of surveys do not differentiate slightly eroded areas from uneroded areas (Fig. 3.9).

3.5.8.2   Moderately Eroded

The cultivated layer typically consists of the surface layer and surrounding layers. Here erosion has so changed the soil that it needs to be managed very differently from the uneroded soil. Standard tillage implements reach the horizontal in most fairly eroded soils or fit conveniently below the single plowed layer (Fig. 3.9).

3.5.8.3   Severely Eroded

In severely eroded soil, the plowing layer generally consists of the underlying material but can also contain a combination of the surface and underlying layers. In some places, shallow or deep gullies are frequently found. Erosion has made a major difference to the soil such that: (1) eroded soil is only appropriate for applications that are significantly less intensive than uneroded soil because it is used on pastures rather than on crops; (2) the eroded soil requires immediate or sustained intensive management; (3) the eroded soil requires immediate or long-term intensive management to be suitable for the same uses as the uneroded soil; (4) productivity is drastically reduced; and (5) engineering operations are more limited than on unmodified soil (Fig. 3.9).

3.5.8.4   Very Severely Eroded

This soil type has an underlying network of slums and is typically of coarse texture and is medium to dark gray. The term 'ravine' generally refers not to an isolated gully but to a complex set of gullies that typically flows into a nearby river formed of deep alluvium. The key systems created by river ravines are wide gully systems (Fig. 3.9).

3.6 Soil Capability

The three factors affecting the evolution of watershed areas are internal soil characteristics, external land features, and environmental factors. Data on the first two aspects are shown on the soil maps created for this research.

Among the essential inherent characteristics of the soil are: efficient soil density; soil texture; reaction; quality of natural matter; and salinity and/or sodium content. Natural surface drainage, slope, flooding, wetness, and gravity are important land features which have a direct effect on soil potential. These factors play a major role in assessing the workability of a specific piece of land for sustainable development in the local environment. A land capability classification is a guide to assessing the suitability of the land for arable, grazing, and forestry. The capability classification is divided into capability classes, capability subclasses, and capability gadgets (Fig. 3.10).

3.7 Planning and Management of Mini-watershed as a Village-Level Water Resource

3.7.1 Preparing a Mini-watershed Project

The study of water resources consists of three principal areas:

a. *Resources*
b. *Environment*
c. *Engineering.*

Each of these areas requires reconnaissance surveys, feasibility studies, planning, design supervision, project management, and training as listed below.

3.7.1.1 Resources

a. Hydrogeological and geological mapping from satellite imagery. Hydrometric network design and monitoring.
b. Assessment of regional surface and groundwater resources as a mini-watershed.
c. Sustainable surface water and groundwater development for irrigation, drinking water, and industrial use within the mini-watershed.
d. Aquifer protection and water conservation.
e. Harvested runoff towards recharging place.
f. Connected uses.
g. Economic evaluation.

3.7.1.2 Environment

a. Hydro environmental impact assessment and western land management.
b. Fallow land and contaminated land studies.
c. Aquifer protection and contaminant migration studies.
d. Hydro-chemical survey.
e. Industrial waste pollution.

3.7.1.3 Engineering

a. Aquifer dewatering; outflow and inflow runoff monitoring within mini-watershed.
b. Hydro-chemical and geotechnical effects on GIS software.
c. Hydrogeological impact on groundwater quality.

3.8 Water Resources Development Plan (WRDP)

An action plan was created for the conservation of water and soil in the watershed. A weighted overlay analysis was used to measure and plan water harvest using various management structures and a zoning map. Weighting and rank were calculated according to the significance of the thematic layer for zoning conservation systems. A control dam, cement bundles, barley package, percolation tank, farm pond, sunken pond, and a loosened boulder system using several thematic layers were suggested as part of the action plan for water and land protection structures (Dongardive et al. 2018).

3.8.1 Groundwater Conditions in the Study Area

Groundwater is generally acceptable for irrigation purposes, except for a few pockets of Risod with more conductivity. Deeper water from the alluvial field cannot be used for irrigation or is of questionable quality. Water analysis from wells in the Akola basaltic terrain shows the water is within the permissible limits for drinking water, excepting Murtizapur, where the saline in the Puma alluvium exceeds the allowable limits (Pande et al. 2014; Moharir et al. 2019).

The water quality is good for drinking and irrigation along the north bank and southern side of the river Puma. Most soils are salty and pH is high. The State Regulatory Authority should specify that the *Zilla Parishad*'s water-use rights are exclusively reserved for drinking water schemes for the villages located in the elementary watershed for the use of runoff for structures of less than 10 mem capability. The right to distribute wastewater rests solely with the district's Central Watershed Management Committee. The *taluka*-level committee determines priorities for the use of the surface runoff. In alluvial areas in Tapi and Puma, instances have been observed of rich farmers with more land and resources digging wells up to 40–50 m deep to obtain water at the expense of poor farmers (Moharir et al. 2020).

3.8.2 Drinking and Irrigation Water Problems

The drinking water problems are:

a. Uneven rainfall distribution across the study region.
b. 80.5% of the region is Deccan Trap rock, which is highly porous and permeable, and almost 48% of its geo-morphologic range has a slope of more than 22%.
c. Uneven surface water quality and amount in all regions. Over-exploitation through widespread creation of open wells in areas where resource issues were present.
d. Sustainability in the summer in several villages with deeper aquifers with a moderate and secure yield for drinking water.
e. Improper construction of tube supply wells for drinking water and lack of evaluation of resources to overcome shortfalls at the implementation stage.

3.8.3 Rooftop Storage

Collection from rooves in areas of settlements is a system for capturing and storing rainwater. The rainwater captured is used for drinking, to a limited extent for farming, or to recharge groundwater.

3.8.4 Storage Tanks

Aquifers usually become saturated in rocky regions during regular rainfall. Thus, unlike for alluvial aquifers, it is not generally necessary to recharge water during the monsoon. Underground storage tanks should therefore be planned according to the need for recharging the drill wells. Until now, much of Maharashtra's industrial area has been supplied from large or medium-sized dams or from continuously flowing rivers. This has continued to provide sufficient storage for growing industries. All industrial units have recently been required by the Indian government to capture and conserve the runoff from their premises' rooftops. The runoff can later be transported in the recharge framework using appropriate filter systems.

3.8.5 Sewage Treatment

Industrial wastewater is processed and re-utilized for agriculture by sewage plants in most developing countries. Some water is treated and reused as potable water in cities instead of being allowed to flow into rivers, thus controlling pollution. The main feature of recharge studies is the reuse of urban and industrial wastewater for recharging groundwater. Wastewater can be recovered by direct injection into wells, water spreading, or irrigation, according to local requirements, water quality, and viability. Wastewater may also be pumped into the water and removed by induced refueling from the wells. Soil aquifer treatment (SAT) is a method for purifying and recovering wastewater and other low-grade water from the soil and the aquifer using physical, chemical, and biological processes. Restored water is collected from wells at a suitable distance to eliminate impurities, unlike groundwater recharging schemes where surface water is applied without any controls. In the SAT system, recovered wastewater flows continuously to the untapped part of the aquifer. All firms in Maharashtra are implementing this type of wastewater treatment program to limit contamination of the groundwater from open wells and reuse the water for irrigation and even for drinking.

3.8.6 Check Dam

A barrier protection dam is a small rock, gravel-bag, sandbag, fiber-sheet, or reusability barrier positioned over a built-in drainage ditch in agricultural land. The check dam decreases the channel's slope, reducing the speed of the water flow, allowing sedimentation and reducing erosion. When a test dam is constructed, an area upstream of the dam is submerged to ensure that the worst places would continue to be overwhelmed. The appropriate location for the building of a check dam can be decided based on the land-use map. From a conservation point of view, the fourth-order stream connecting to the mainstream was chosen for a dam site.

A check dam may be appropriate in the following situations:

- To promote sedimentation.
- To prevent erosion in small seasonal canals or temporary swales by decreasing the channel flow speed.
- In a small open channel that drains 10 acres or less.
- In a steep channel where stormwater runoff velocities exceed 5 ft/s.
- When grass filling and drainage ditches or canals are to be created.
- In temporary ditches where erosion-resistant lining is not required for the a limited time.

The check dam is not to be used for extended flows in live streams or channels. Nor is it suitable for canals that drain over 10 acres, or in grass-lined channels, as planting will destroy vegetation unless erosion is expected. It encourages the capture of sediments that may be re-suspended or removed during subsequent storms.

3.8.7 Cement Nala Bund

This is a low weir without a drainage canal, but it provides lift irrigation and agriculture under the wells in the field. It also helps to restore water sources that are exhausted by wells. Tube wells are usually supported on little streams or nala with a continuous flow, especially during the rabi season. Terrestrial soil areas where soil depth is greater than 100 cm are proposed for the structural cement nala band (Plate 3.1).

The following conditions should be observed:

a. The stream should be straight both upstream and downstream.
b. The nala bed slope should not be more than 3%.
c. The nala width should be less than 30 m.
d. The nala should have a stable bank on either side.
e. The rock level below the bed level should not be more than half the height of the structure above the ground level.
f. The structure should not in any way lead to water spreading into nearby agricultural fields.

3.8.8 Continuous Contour Trench (CCT)

In steeply sloping or hilly areas with heavy precipitation, continuous contour trenches are built. Runoff and soil degradation can be minimized in steep pathways. The continuous contour trenches are suitable on slopes of 3–5% with very shallow to shallow soil (>10 cm). Soil and water management trenches are dug into the slopes and around the contour lines.

3.8.9 Loose Boulder Structure (LBS)

Loose boulder structures are appropriate on slopes of 5–10%, with gravel sandy clay and low ground depth (>10 cm) on agricultural cropland. They are suitable where there are multiple stones and gullies of small to medium-sized drainage areas in moderate pitfalls. Their flexibility and weight help to maintain contact with the ground. Flat stones are the best material to use (Plate 3.2).

3.8.10   Earthen Nala Bund (ENB)

An earthen nala bund is suggested on forested slopes of 1–3%, where the soil is very deep (>100 cm). The dam is built over a rivulet and holds the water for a period, allowing it to percolate and gradually raising the level of subsoil water surrounding it. The site requires a narrow valley for soil filling, with a slope of no more than 3%. The width of the nala should be more than 5 meters. The system will not contribute to water spreading in the surrounding farmland. The fundamental strata should be able to bear the structure's load.

3.8.11   Sunken Pond

These are constructed to create more storage by recharging the groundwater through percolation. They are constructed in suitable places across the nala. The site should be relatively flat, the slope not exceeding 2%. The catchment of the structure should be above 40 ha. An emergency spillway, preferably in hard rock, may be provided alongside a percolation tank bund.

3.8.12   Percolation Tank

Percolation tanks installed across the nala generate additional soil water storage and aquifer recharging by percolation. They also provide drinking water to cattle and in many cases also to the human population. Percolation tanks should be constructed preferably on second-to third-order streams, and located on highly fractured and weathered rocks with downstream lateral continuity.

Appropriate sites, ideally in hard rock, are relatively flat with a slope of 2%. The system catchment covers 40 ha. An emergency spillway can be provided to avoid binding of the percolation tank. In many cases the structure also provides drinking water for cattle and people.

3.8.13   Graded Bund

In addition to soil protection, safe disposal of excess moisture is of key importance in areas of higher rainfall. In this region, excess moisture should be funneled into graded sites after it has been allowed to soak into the soil to the necessary depth. Graded bunds conserve excess water in a natural vegetated waterway. They are designed to channel water at non-erosive speeds, flooding the ground up to root level. Graded bunds may also offer advantages in low rainfall areas as they eliminate the need for side bunds. Length, grade, and parts of the bund must, therefore, be carefully calibrated according to the nature and quantity of rainfall and soil type (Pande 2014b).

3.8.14 Canal Control

Between 1 and 3% of the farmland is covered by an irrigation network of canals. The soil type is clay and the depth ranges from shallow to very deep (>100 cm).

3.8.15 Pond

A reservoir pond is suggested for rainwater storage in a specific agricultural area and for irrigation purposes. This is suitable on slopes varying from 1% to 3% with clay soil and depth of small to very deep (>100 cm).

3.9 Water Resource Development

There is a variety of alluvial and Deccan Trap exposures in the study area. Water resources in the area can be divided into three distinct groups: (i) good, in which the electrical conductivity (EC) value is between 250 and 750; (ii) medium, with EC values of 750–2000; and (iii) doubtful, where the EC value is between 2000 and 3000. The alluvial zone, with EC values reducing towards the south, has doubtful water quality that is not suitable for drinking. Interestingly, none of the analyzed samples shows EC > 3000, which means that the salinity rate of the alluvial region of the study area is in the lower range and can be eliminated by effective drainage methods and also by rainfall pumping of salty water into the Man River (Khadri ans Chaitanya 2015a, b).

Variations in the water level are probably due to distinct changes in precipitation and the indiscriminate extraction of groundwater in the area. Extensive climate and weather data obtained over ten years from observation wells were used to understand the complexity of the groundwater system and identify suitable sites for groundwater production. Numerous measures are proposed to boost the capacity and volatility of groundwater levels in the area with a focus on environmental management (Prakasam 2010).

The region is distinguished by the presence of an alluvial field that shows saline flows in the south and basaltic lava flows in the north. Tapped with dug wells, small, unconfined aquifers are the biggest groundwater producers in the region (Pande et al. 2016). Groundwater salinity is largely regulated by lithology, with high carbonate- and evaporite-regulating salinity showing different chemical relations between soil and rock lithology. Water chemistry depends mainly on minor alteration of the host rock. In this watershed area there is good scope for hydrological investigation and environmental protection measures. Investigations provide useful clues for the assessment of groundwater potential in the region. They can also assist in the reconstruction of groundwater routes in a particular watershed, which are of use for optimal soil and water planning and management. In this area, groundwater replenishment is determined by topographical characteristics, including thickness of the weathered zone, and soil and soil strata permeability in the aeration region (Pande et al. 2017). Red boles reveal seven lava streams isolated from each other in the region. Every part of the flow forms a different unit depending on the porosity and permeability of the flow units. The water content of various lava flows depends on the form and composition of the eruption. Broad sections having low porosity are not beneficial to groundwater. In comparison, vesicular and amygdaloidal horizons in lava flows with interconnected and evenly distributed vesicles are extremely porous and permeable. Their contribution to the groundwater potential depends on varying weather conditions. Occasionally porosity can be formed by closely spaced interconnecting joints between massive horizons. On basaltic soil, the size and number of vesicles, degree of weathering and joint patterns are mainly responsible for water productivity and water supply potential. Heavily weathered vesicular and amygdaloidal basalt regions are therefore good producers of soil water (Moharir and Pande 2014a, b).

Thus, water chemistry is consistent across the area of the traps, while the alluvial region shows some changes in chemistry due to salinity. Variations in pH and total dissolved solids (TDS) are influenced by lithology and environment (Pande and Kanak 2018c). In downstream temperate areas, chemical weathering plays an important role in water chemistry regulation. Low water quality not suitable for drinking and irrigation is observed in the northern saline region (Khadri et al. 2013; Pande et al. 2019a). Decent to excellent groundwater quality is found, however, in the area of the traps. Within our study area, the maximum amount of water is found in the highly divided, weathered, and joined horizons of the Deccan Traps. There are also many aquifers that show both productive and unproductive zones due to the presence of laterally varying massive and vesicular units (Khadri and Chaitanya 2015c). Groundwater levels fluctuate between pre-monsoon and post-monsoon. Water depth studies indicate variations in groundwater table replenishment according to irrigation methods. Based on field characteristics, specific mineral levels, textural characteristics, and distribution of geochemical signatures, the existence of ten lavic flows belonging to the Atali, Lokhanda, and Amdapur formations has been established. A detailed geological map of the study area depicting the lava sequence has been prepared.

3.10 Conclusion and Recommendation

This chapter has shown the importance of thematic mapping for novel models of sustainable watershed development in the semi-arid region. Watersheds are biophysical systems with structures designed to create and maintain land and water resources on a sustainable basis. The significance of an IMM solution for addressing ground, water, and hydraulic extremes should be understood by the reader.

References

Dongardive MB, Patode RS, Nagdeve MB, Gabhane VV, Pande CB (2018) Water resources planning for the micro watersheds using geospatial techniques, IJCS 6(5):2950–2955

Khadri SFR, Chaitanya P, Kanak M (2013) Groundwater quality mapping of PTU-1 Watershed in Akola district of Maharashtra India using geographic information system techniques. Int J Sci Eng Res 4(9)

Khadri SFR, Chaitanya P (2015a) Remote sensing based hydro-geomorphological mapping of Mahesh river basin, akola and Buldhana Districts, Maharashtra, India—effects for water resource evaluation and management. Int J Geol Earth Environ Sci 5(2):178–187

Khadri SFR, Chaitanya P (2015b) Groundwater quality Mapping using hydrogen chemistry and Geostatistical analyst of Mahesh River Basin, Akola and Buldhana District, Maharashtra, India. Int J Res (IJR) 2(10)

Khadri SFR, Chaitanya P (2015c) Analysis of hydro-geochemical characteristics of groundwater quality parameters in hard rocks of Mahesh River Basin, Akola, and Buldhana Dist Maharashtra. India Using Geo-Inf Techn Am J Geophys Geochem Geosyst 1(3):105–114

Khadri SFR, Chaitanya P (2016) Ground water flow modeling for calibrating steady State Using MODFLOW software—a case study of Mahesh River Basin, India. Model Earth Syst Environ 2(1):2. ISSN 2363-6203

Khadri SFR, Pande C (2014) Hypsometric Analysis of the Mahesh River Basin In Akola and Buldhana Districts Using Remote Sensing & GIS Technology. Int J Golden Res Thoughts 3(8)

Khadri SFR, Pande C (2015) Remote Sensing and GIS Applications of Linament mapping of Mahesh River Basin, Akola & Buldhana District, Maharashtra, India Using Multispectral Satellite Data. Int J Res (IJR) 2(10)

Kokate NR, Moharir KN, Pande CB (2014) Morphometric analysis of Ural Khurd Nala watershed in Akola District of Maharashtra, India: Using remote sensing and geographic information system (GIS) techniques. Int J Res (IJR) 1(11)

Moharir KN, Pande CB (2014a) Analysis of Morphometric Parameters Using Remote-Sensing and GIS Techniques. In: The Lonar Nala in Akola District, Maharashtra, India. Int J Technol Res Eng 1(10)

Moharir K, Pande C (2014b) Change Detection in Forest/ Non-Forest Cover Mapping Using Remote Sensing and GIS Techniques of Patur Taluka in Akola District, Maharashtra. Int J Res (IJR) 1(10)

Moharir K, Chaitanya P, Sanjay P (2017) Inverse modeling of Aquifer parameters in basaltic rock with the help of pumping test method using MODFLOW software. Geosci Front, May 2017

Moharir K, Pande C, Singh S, Choudhari P, Rawat K, Jeyakumar L (2019) Spatial interpolation approach-based appraisal of groundwater quality of arid regions. Aqua J 68(6):431–447

Moharir KN, Pande CB, Singh SK, Del Rio RA (2020) Evaluation of Analytical Methods to Study Aquifer Properties with Pumping Test in Deccan Basalt Region of the Morna River Basin in Akola District of Maharashtra in India, Groundwater Hydrology, Intec open Publication, UK. https://doi.org/10.5772/intechopen.84632

Pande C (2014b) Change detection in land use/land cover in Akola Taluka using remote sensing and GIS Technique. Int J Res (IJR) 1(8)

Pande C, Moharir K (2014) Analysis of land use/land cover changes using remote sensing data and GIS techniques of Patur Taluka, Maharashtra, India. Int J Pure Appl Res Eng Technol 2(12):85–92

Pande CB, Kanak M (2015) GIS-based quantitative morphometric analysis and its consequences: a case study from Shanur River Basin, Maharashtra India. Appl Water Sci 7(2)

Pande CB, Kanak M (2018c) Spatial analysis of groundwater quality mapping in hard rock area in the Akola and Buldhana districts of Maharashtra, India. Appl Water Sci 8(4):1–17

Pande CB, Moharir KN, Khadri SFR, Sanjay P (2018b) Study of land use classification in the arid region using multispectral satellite images. Appl Water Sci 8(5):1–11

Pande CB, Moharir K, Rajeshwari P (2018a) Assessment of morphometric and hypsometric study for watershed development using spatial technology—a case study of Wardha river basin in Maharashtra, India. Int J River Basin Manag https://doi.org/10.1080/15715124.2018.1505737

Pande CB, Khadri SFR, Moharir KN, Patode RS (2017) Assessment of groundwater potential zonation of Mahesh River basin Akola and Buldhana districts, Maharashtra, India using remote sensing and GIS techniques. Sustain Water Resour Manag. https://doi.org/10.1007/s40899-017-0193-5

Pande CB, Moharir KN, Deshkar BS (2014) Morphometric Analysis of Pt-7 Sub-watershed Using Remote Sensing And GIS Technology. In: Akola District, Maharashtra. Int J Pure Appl Res Eng Technol (IJPRET) 2(11):8–25

Pande CB, Moharir KN, Singh SK, Dzwairo B (2019a) Groundwater evaluation for drinking purposes using statistical index: study of Akola and Buldhana districts of Maharashtra, India. Environ Dev Sustain (A Multidisciplinary Approach to the Theory and Practice of Sustainable Development). https://doi.org/10.1007/s10668-019-00531-0

Pande C, Pande B, Moharir KN, Singh SK, Varade AM (2019b) An integrated approach to delineate the groundwater potential zones in Devdari watershed area of Akola district, Maharashtra, Central India, Environment, Development and Sustainability. https://doi.org/10.1007/s10668-019-004 09-1

Patel M, Dhiraj K (2019) Water resource management plan of a micro-watershed using geospatial techniques. Int J Curr Microbiol App Sci 8(02):270–277. https://doi.org/10.20546/ijcmas.2019.802.032

Patode RS, Nagdeve MB, Pande CB, Moharir KN (2017a) Land use and land cover changes in Devdari Watershed Tq. Patur, Distt. Akola, of Vidarbha Region in Maharashtra. Trends Biosci 10 (8)

Patode RS, Pande CB, Nagdeve MB, Moharir KN, Wankhade RM (2017b) Planning of conservation measures for watershed management and development by using geospatial technology—a case study of Patur Watershed in Akola District of Maharashtra. Curr World Environ 12(3)

Prakasam C (2010) Land use and land cover change detection through remote sensing approach: a case study of Kodaikanal taluk Tamilnadu. Int J Geomatics Geosci 1(2):150–158

Reddy, Mahender D, Patode RS, Nagdeve MB, Satpute GU, Pande CB (2017) Land use mapping of The Warkhed micro-watershed with geo-spatial technology. Contemp Res India 7(3). ISSN 2231-2137

Sarma VVLN, Murali Krishna G, Hema Malini B, Nageswara Rao K (2001) Land use land cover change detection through remote sensing and its climatic implications in the Godavari delta region. J Indian Soc Remote Sens 29(1&2)

Chapter 4
Sustainable Watershed Development Planning

Abstract In semi-arid regions, sustainable water change will help boost quality of soil, and drinking and irrigation water, and this chapter proposes a strategy for the development of land and water resources for sustainable watershed management. The main aim of the study is to directly impact the effects of climate change on groundwater, sustainable farming, and surface water. The planning of artificial groundwater recharge sites relies on zoned mapping. Remote sensing and GIS software is used to demarcate ground and artificial refilling sites. Groundwater and artificial recharge planning maps were created and weighted overlay evaluation methods used for integrated data such as soil types, land capacities, soil pitches, land use/land cover and soil drainage. The maps were classified by Arc GIS software 10.1 with numerical values 1–10.

Keywords Remote sensing · GIS · Soil types · Semi-arid area

4.1 Introduction

In recent years, the role of individual artificial water replenishment planning has become better understood in India. Scientists and NGOs have concentrated on statistical modeling using geological, geomorphological, subsurface, geological, and water-level fluctuation data, and remote sensing and GIS methodology, to extend artificial watershed replenishment systems (Anbazhagan and Ramasamy 2005). GIS and land use are natural companions as both of them deal with spatial details.

With rapid urbanization and increasing population in urban areas, the demand for water is sure to increase and can also contribute to the possible depletion of groundwater reservoirs. Restoring water levels in the depleted groundwater reservoirs (aquifers) can be achieved by artificial recharging. The central and state governments spend a great deal of money on the construction of artificial systems such as percolation ponds and test dams, and scientific research based on specific parameters is required to choose suitable locations, define site-specific processes, estimate surface runoff, and prioritize synthetic recharging activities (FAO 2003). Subsurface reservoirs are very desirable alternatives that are technically feasible for storing excess

C. B. Pande, *Sustainable Watershed Development*,
SpringerBriefs in Water Science and Technology,
https://doi.org/10.1007/978-3-030-47244-3_4

monsoon runoff. Water-spreading strategies include over-irrigation, the construction of basins, the use of engineering techniques, or mechanical modifications to natural conditions such as improving a stream channel (Reddy et al. 2017). In areas where the aquifer has been exhausted by over-development, artificial recharge techniques typically lead to improved sustainable yield and the recovery of surplus surface water for future needs (Moharir et al. 2017; Pande 2014; Patode et al. 2017).

Artificial recharge involves the movement of natural surface water through underground structures. Recharge can be either direct or indirect. For direct recovery, injection wells pump water into an aquifer. The injected water is treated to ensure the area around the injection is no longer clogged. Indirect drainage involves spreading surface water over land so that the water infiltrates the aquifer through the vadose region, the unsaturated layer above the water table.

The goal of this study covering the Akola and Buldhana districts in the Maharashtra semi-arid region was to identify important issues concerning management of groundwater with the aid of state-of-the-art technology (Pande and Moharir 2018c). The study provides unique field and laboratory knowledge with GIS analysis filling fundamental gaps in knowledge associated with the problem of salinity. The research area consists of a variety of erosional surfaces, like terraces in phase. The basin has an aerial length of 328.25 km and is located in two administrative *talukas*, Balapur and Khamgaon. The basin is filled geologically mainly by Deccan rock types. Most of the cultivated area is irrigated and the soil improved, with the groundwater provided from open and viable wells. In this environment, water tank replenishment by synthetic refilling is necessary.

Preparations for synthetic recharging were made, based on remote sensing data, after the magnitude of the accessible runoff at each watershed had been estimated. The final stage involved the integrated study of multidisciplinary information sets to assemble composite information that is able to answer a variety of queries.

4.2 Priority Groundwater Recharge Zone Mapping

The research uses remote sensing and GIS technology to focus on artificial recharge sites and runoff evaluation methods. Satellite data and aerial photographs from Indian Remote Sensing Satellite (IRS 1A) with Linear Imaging Self-scanning Sensors (LISS II) were used to generate many digitally processed thematic maps with geological and geomorphological parameters. The thematic maps developed by the ARC GIS program show layers such as lithology, lineament, lineament density, composition, rivers, and geomorphology. To assess geological, geomorphological, and hydrological data, the Indian Survey (SOI) has also been used for topographical maps, geophysical resistivities, and area investigations.

Base vector files were created to enable layer evaluation. Following the preparation of these base layers, the drainage layer was converted to a drainage density layer and the lineament layer to a lineament density layer. Reasonable weights were allocated to different land use/cover groups for each thematic map. The weightings were applied to the reclassified DEM based entirely on their topographical suitability for water recharge (Table 8.5). The weightings covered six geological classes, eight geomorphological classes, four soil classes, six drainage densities, and five classes of lineament density. All the allocated weights were averaged by dividing the individual weights for the different layers according to the total weight. These normalized weights were added to the weights already assigned to the thematic layers to give total weights. All the vector layers were converted to raster layers based on the assigned full weights to identify the suitability of the location for groundwater recharge. A site suitability map was developed, reclassified into distinctive groups after the inclusion of all layers: priority (highly suitable), second priority (moderately suitable), and not currently suitable (lowest priority area) (Fig. 8.12).

The quantity of runoff available in every watershed and the artificial recharge planning were carried out using remote sensing data and GIS software. Integration and prioritization of natural resource information was accomplished in the ARC GIS setting by thematic mapping and statistical analysis.

Several related facets of the mapping were evaluated for multi-thematic overlay analysis based on satellite data. For example, IRS LISS-III satellite information was used for the planning of a plane erosion by superposing specific land-use and land cover, geomorphology, soil profile, and streams. Thematic layers and geological data are described based on information provided by the National Office of Land Use Planning and Soil Survey (NBSS and LUP) and the Indian Geological Survey (GSI) (Pande et al. 2018b) (Tables 4.1 and 4.2).

Table 4.1 Geological classes including structure

S. N.	Lithology/structural features	Weighting
1	Massive basalt	1
2	Vesicular basalt	3
3	Amygdaloidal basalt	4
4	Compact basalt	2
5	Fault	5
6	Joint	4

Table 4.2 Geomorphological classes

S. N.	Lithology/structural features	Weighting
1	Plateau	4
2	Alluvial area	6
3	Habitations mask	2
4	Waterbody	1

4.3 Artificial Recharge Site Selection

In water shortfall areas, artificial recharge is a process for extending the natural movement of surface water to underground structures. It is achieved by designing facilities for infiltration or by inducing regeneration from groundwater bodies. In countries like India, artificial recharge planning is critical for the sustainable development of the watershed. The overall success of these activities can be tremendously improved if they are conducted with adequate scientific preparation. Remote sensing and GIS data can be a very useful method of planning for synthetic charging structures, although until now they have not been widely used in India.

In the present research, site selection is based solely on hydrogeological perspectives, and engineering considerations are not included. The following information is usually relevant for selection of an artificial recharge site: recharge water source; suitable geological formation; thickness and permeability of the tissue surrounding the geological formation; proximity of the possible recharge site to the gloomy cone of a fantastic property; and differences in water levels. An attempt was made to classify artificial recharge sites according to elements such as lineaments, hydrogeomorphic and hydrogeological features, land use, and drainage capacity. Weights have been assigned to the thematic maps which determine their relative percentage scores. Overlay analysis of separate thematic layers was performed to determine areas and sites suitable for synthetic recharge. Resource-themed mapping data were extracted from LISS-III satellite imagery from February 2013 with a spatial decision of 23.5 m. Topographical survey sheets, geological maps, and a soil map of the area were also used. Lineament density and hydrological soil classification were defined based on infiltration rate, texture, depth, drainage, and water infiltration capacity. Specific morphometric parameters for the basin were determined using common methods.

A system based on remote sensing and GIS is very useful in assessing suitability for artificial recharge sites in the sub-watershed. The first task was to identify the factors which would facilitate the recharge. The current artificial recharge network in the region has been studied in terms of hydro-geomorphology, topography, and well water levels. Based on these findings, a set of rules was developed to demarcate the most appropriate zones and also to define the exact artificial recharge sites (Pande and Moharir 2015). The thematic maps required for site suitability analysis using the weighted indexing method were: (a) geology; (b) geomorphology; and (c) land use and land cover. The higher-value groups contain the most desirable zones for artificial recharge structures (Fig. 4.1).

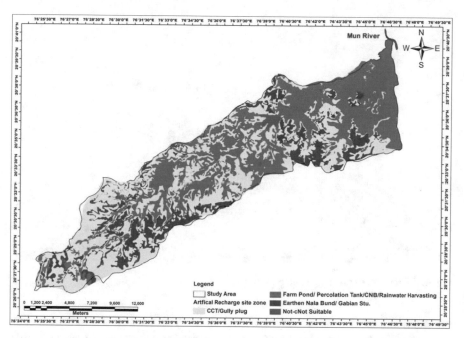

Fig. 4.1 Artificial recharge site zones map

4.3.1 Identification of Priority Groundwater Recharge Sites Using Remote Sensing and GIS Technology

Remote sensing and GIS technologies were used to delineate the sites for groundwater recharge. Remote sensing plays a huge role in the field of development of hydrology and water resources, offering multi-spectral, spatial, and sensor knowledge of surfaces worldwide. One of the main benefits of remote sensing technology for hydrological investigations and monitoring is its capacity for providing information on spatial and temporal formats (Pande et al. 2018a), which are important for effective investigation, prediction, and validation of maps of groundwater recharge sites for the semi-arid environment. However, remote sensing involves significant amounts of spatial information management which needs an environmentally friendly system to handle it. The study promoted the use of advanced geospatial techniques to produce spatial knowledge of potential recharge sites and assess their suitability.

Laterite is the main aquifer in the Mahesh river basin. In this aquifer, groundwater exists in water table conditions and is extensively used for domestic and irrigation purposes (Pande et al. 2019a; Khadri et al. 2015a, b). Depositions of valley filling and weathered basalts often locally reflect significant unconfined aquifers. The small aquifers made of iron ore and broken schistose rocks are, however, not often used. Groundwater in the unrestricted aquifers is under 6 m below ground level. While there is a widespread upward drive in the water table due to rainfall recharge during

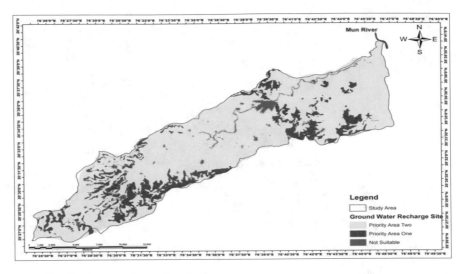

Fig. 4.2 Groundwater recharge site priority area

the monsoon season, the water stages fall abruptly as soon as the monsoon ends, suggesting the unrestricted aquifers along topographic slopes are relatively drainable. The fluctuation of the water table is less than 2 m across a vast watershed area, resulting in a low amount of complex groundwater supply in the watershed (Khadri and Pande 2016b, c). Two groundwater domains turn up at two different topographic levels in the watershed, separated by the Western Ghats escarpment. The tributaries are affluent and acquire base flow throughout the year from the two groundwater domains (Fig. 4.2).

4.3.2   Groundwater Resource Evaluation

A critical problem in groundwater production is the observable measurement of the aquifer recharge. Groundwater recharge assessment includes a proper understanding of recharge and discharge procedures and their interrelationship with geological, geomorphological, soil, land use and climatic factors affecting the watershed region. Several strategies are used for quantitative groundwater recharge comparison: (a) groundwater fluctuation degree and basic yield technique; (b) rainfall infiltration technique; and (c) soil moisture balance system. The groundwater degree fluctuation and specific yield approaches are used in the current study to quantitatively estimate groundwater recharge in the semi-arid territory. Traditional groundwater recharge assessment techniques are limitations, despite their ease and broad applicability in various hydrogeological systems (Khadri and Pande 2014a, b).

Groundwater movement is governed by natural boundaries. In traditional approaches such as the water-level fluctuation method, the average fluctuation of the water level is measured as part of the field analysis. Spatial variation in the charging factors is not taken into account. In the remote sensing and GIS-based approach, attention is given to the spatial distribution of variables, so that a data layer for the whole of a watershed can be planned. Remote sensing records also provide the most accurate ground data, minimizing fieldwork. Seasonal data are necessary for recharge estimation (Khadri et al. 2013).

4.3.3 Weighted Index Overlay for Identification of Groundwater Potential Zones

Weighted index overlay analysis (WIOA) is a simple and straightforward technique for a joint analysis of multi-class maps. The usefulness of this approach lies in the ability to include the element of human judgment in the study. The WIOA method contemplates the parameters and their sub-groups. A simple weighted overlay technique has no trend scale. Measurement criteria are defined, and importance is assigned to each parameter. Determination of all types of weighting is the most integral part of integrated analysis since the output is usually organized with the correct weighting (Pande et al. 2019b). Looking at the relative value leads to a clearer description of real conditions in the field. Prospective groundwater zones were delineated considering the hydro-geomorphic conditions of nearby weighted indexing, taking into account five important parameters: geomorphology; geology; soil; land use/land cover; and water level (Fig. 4.3 and Table 4.3).

4.3.4 Runoff Estimation

Another important element in artificial recharge work is the estimation of water available as runoff. In our study, following the estimate of runoff, areas for artificial regeneration were prioritized based entirely on the available water and aquifer measurements at each watershed. To estimate runoff in the watershed areas, a database of the following is required: aerial coverage of different land use and land cover, hydrological soil group, and rainfall.

4.4 Groundwater Availability

During the dry season (pre-monsoon) the thickness of a water column that has been drilled through the entire thickness of the unconfined aquifer is a good measure of the availability of groundwater in a given vicinity. Wells are graded based on the thickness of the water column, meaning that the wells in the Mahesh river basin were

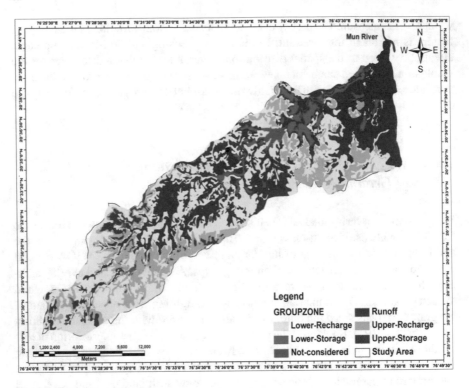

Fig. 4.3 Recharge zone map

drilled through the entire thickness of the unconfined aquifer (Moharir et al. 2020). In two consecutive years of water observations for the post-monsoon (November) and pre-monsoon (May) seasons, a specific value for the water column has been estimated by subtracting the depth of groundwater below ground from the overall depth of the property underground (Patode et al. 2016). If the water column remains 50% above the post-monsoon water level then the water availability is classed as good. Between 25ô and 50% it is average, and less than 25% is bad (Fig. 4.3).

4.5   Groundwater Potential Zones

The drainage basin morphology is an important aspect of geomorphic research. Many hydro-geomorphological features, together with their geological parameters, can be observed and evaluated using remote sensing techniques. This is very useful for the preparation of integrated hydro-geomorphological maps for groundwater concentrations (Khadri et al. 2016a). Using satellite imagery visual analysis, topographical maps, and topic search, the area can be divided into numerous hydro-geomorphic units demarcating workable construction zones (Pande et al. 2017).

Table 4.3 Ratings for parameters on a scale of 1–10

Parameter	Ranks (in %)	Classes/units	Weight
Hydro-geomorphology	25	Moderate	6
		Moderately dissected	5
		Slightly dissected	3
		Undissected	1
		Weathered	4
Land use and land cover	20	Agriculture	7
		Wastelands	4
		Forest	2
		Built-up land	1
		Water bodies	7
Slope	15	0–3.26	4
Digital elevation model	15	260–598	5
Water level	25	6.1–7.2	3
		7.2–8	2
		8–8.7	3
		8.7–9.5	4
		9.5–11	4
		11–12	2
		12–14	2
		14–17	2

Hydro-geomorphological units such as alluvial plains and filled valleys are the most suitable exploration and production zones in groundwater analysis (Khadri et al. 2015c, d) and are classified as good to very good. Upland (deep, moderate, shallow) regions are identified as moderately favorable, and denuded upland regions with low lineament density are the least favorable areas for groundwater exploration and development. Figure 4.4 shows that the southern part of the basin has excellent groundwater capacity relative to the upper-middle and northeastern part, and this has also been tested in the field. This knowledge is very useful for the further production of groundwater within the study area (Biswas et al. 2009; Kale and Kulkarni 1993).

Once all thematic maps had been integrated, the groundwater potential map was divided into several zones. The map clearly shows that the alluvial plain, which consists of sand, silt, and clay with a near-level slope and very low drainage density, has very good potential and is a highly promising area for groundwater extraction along with lineaments. The structural, denuded, and residual hills are potentially low to extremely poor groundwater zones (Fig. 4.4) but they act as runoff areas. Lineaments, especially joints, fractures, and their intersection, enhance the potential of a hydrogeomorphic unit. This potential groundwater map provides the basis for future exploration (Pande et al. 2019b).

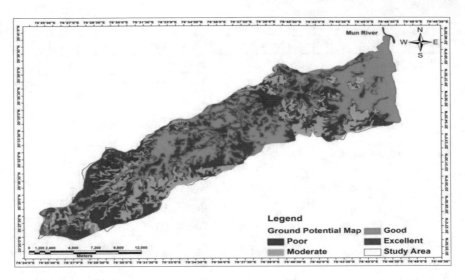

Fig. 4.4 Groundwater potential zones map

Table 4.4 Groundwater potential area classes and area covered

Groundwater potential	Area (km^2)	Percentage
Poor	181.37	55.25
Moderate	109.24	33.28
Good	34.53	10.52
Excellent	3.0	0.91
Total area	328.25	100

4.5.1 Excellent

The excellent zone includes valley fill, flood plains, and low-lying areas as well as lineament intersections such as cracks, fractures, and joints. It usually includes areas where unconsolidated sediments have been deposited, such as gravel, sand, silt, and clayey sand. These have high water retention potential, as they allow maximum percolation between the grains due to their maximum pore space. This zone covers an area of approximately 3 km^2 and forms 0.91% of the study area (Fig. 4.4 and Table 4.4).

4.5.2 Good

All the remaining controlled geological structures fall into the strong potential category. Many areas with low-lying and gentle slopes are also included. In general, sandstone is capable of storing and transmitting water through interstices and pore spaces between the grains and is considered ideal for aquifers. Areas of exposed sandstone also come into this category. This region covers an area of around 34.53 km^2, 10.52% of the study area (Fig. 4.4 and Table 4.4).

4.5.3 Moderate

This zone comprises mainly areas where the recharge state and the water-yielding capacity of the materials underlying it is neither suitable nor bad. Topographically, it occupies the smooth, gently sloping hilltops. While the lithology includes good water-bearing rock formations such as sandstone, the potential is reduced by the slope, where maximum runoff is present. The moderate zone typically includes low water-bearing rock formations such as silty shale, which in turn is distinguished by the presence of secondary structures within it. The moderate zone is evenly distributed within the study area, covers an area of 109.24 km^2, and occupies 33.28% of the total study area (Fig. 4.4 and Table 4.4).

4.5.4 Poor

This zone is located primarily in the elevated regions. Much of the precipitation flows out as surface runoff in the high relief regions, leading to poor conditions for infiltration under the ground. The groundwater yield is therefore usually deemed low unless the elevated areas are traversed by geological systems, and possess high drainage density and sufficient water-bearing rock formations. The poor zone is mostly scattered along the ridges and covers the majority (55.25%) of the study area.

4.6 Conclusion

Thematic mapping of the study area and its inherent characteristics was performed using geospatial techniques. Recently collected groundwater data from specific wells identifies an increase in groundwater depletion and shows the water table has a seasonally declining pattern. Dendritic to sub-dendritic drainage patterns with a moderately dense drainage texture were found in the sub-watersheds covering the study region. High bifurcation ratios indicated good structural drainage power and strong headward erosion. Overlay analysis of various thematic layers (geomorphology, slope, drainage, drainage depth, and land use) culminated in the production of a final map giving a general understanding of the area's potential for groundwater and showing suitable groundwater recharge areas. The incorporation of the thematic layers used a model developed using GIS techniques.

Groundwater is a precious finite resource. Increasing population, urbanization, and the expansion of agriculture have, over the years, led to unscientific groundwater exploitation creating conditions of water stress. A cost-effective and time-efficient technique for proper groundwater resource assessment and management planning is needed. Groundwater planning software involves a large amount of data coming from different sources. As successfully demonstrated in this report, integrated remote sensing and GIS can provide the proper forum for decision taking on groundwater studies based on convergent analysis of large volumes of multidisciplinary data. The groundwater potential map is a systematic project that takes into account major control factors that affect water yield, artificial recharge location, and groundwater quality. The map is important as the basis for groundwater exploration planning and execution.

The following conclusions may be drawn from the study.

a. The role of remote sensing and GIS-based methods of groundwater resource evaluation were developed and demonstrated in the study.
b. The study shows that recharge sites located on a gentle slope and lower-order streams are likely to provide artificial recharge to a larger region.
c. In selecting suitable sites for artificial recharge, a model combining geology, land-use ground cover, geomorphology, contour, soil, and digital elevation was found to be very useful.
d. The change in land use is primarily due to hydrological factors.

References

Anbazhagan S, Ramasamy SM (2005) Evaluation of areas for artificial groundwater recharge in Ayyar basin, Tamil Nadu, India through statistical terrain analysis. J Geol Soci India
Biswas AK, Tortajada C, Izquierdo R (eds) (2009) Water management in 2020 and beyond. Springer, Berlin. https://doi.org/10.1007/s10668-019-00409-1
FAO (2003) World agriculture: towards 2015/2030. FAO An FAO perspective, Rome
Kale VS, Kulkarni H (1993) IRS-1A and landsat data in mapping deccan trap flows around Pune, India: implications on hydro-geological modelling. Int Arch Photogramm Remote Sens 29:429–435
Khadri SFR Pande C, Moharir K (2013) Groundwater quality mapping of PTU-1 Watershed in Akola district of Maharashtra India using geographic information system techniques. Int J Scient Eng Res 4(9)
Khadri SFR, Pande CB (2014a) Remote sensing and GIS Applications of geomorphological mapping of Mahesh River Basin, Akola & Buldhana Districts, Maharashtra, India using Multispectral Satellite Data. Indian Streams Res J 4(5)
Khadri SFR, Pande CB (2014b) Morphometric analysis of Mahesh river basin exposed in Akola and Buldhana Districts, Maharashtra, India Using Remote Sensing &GIS Techniques. Int J Golden Res Thoughts 3(1), ISSN 2231-5063
Khadri SFR, Pande Chaitanya (2015a) Analysis of Hydro-geochemical characteristics of groundwater quality parameters in hard rocks of Mahesh River Basin, Akola, and Buldhana Dist Maharashtra, India using geo-informatics techniques. Am J Geophys Geochem Geosyst 1(3):105–114

Khadri SFR, Pande Chaitanya (2015b) Remote sensing based hydro-geomorphological mapping of Mahesh river Basin, Akola, and Buldhana Districts, Maharashtra, India—effects for water resource evaluation and management. Int J Geol Earth Environ Sci 5(2):178–187

Khadri SFR, Pande C (2015c) Remote sensing and GIS applications of Linament mapping of Mahesh River Basin, Akola & Buldhana District, Maharashtra, India Using Multispectral Satellite Data. Int J Res (IJR) 2(10)

Khadri SFR, Pande C (2016a) Ground water flow modeling for calibrating steady state using MODFLOW software—a case Study of Mahesh River Basin, India. Model Earth Syst Enviro 2 (1), ISSN 2363-6203; Khadri SFR, Pande C (2016b) GIS-based analysis of groundwater variation in Mahesh River Basin, Akola and Buldhana Districts, Maharashtra, India. IJPRET 4(9):127–136

Khadri SFR, Pande C (2016b) Geo-environmental resource management in Mahesh River Basin in Akola and Buldhana Districts, Maharashtra Using Remote Sensing and GIS Techniques. IJPRET 4(9):151–163

Moharir K, Pande C, Patil S (2017) Inverse modeling of Aquifer parameters in basaltic rock with the help of pumping test method using MODFLOW software, Geoscience Frontiers, May 2017, 1–13

Moharir KN, Pande CB, Singh SK, Del Rio RA (2020) Evaluation of analytical methods to study aquifer properties with pumping test in deccan basalt region of the morna river basin in akola district of Maharashtra in India. In: Groundwater Hydrology, Intec open Publication, UK. https://doi.org/10.5772/intechopen.84632

Pande C (2014) Change detection in land use/land cover in Akola Taluka using remote sensing and GIS technique. Int J Res (IJR) 1(8)

Pande CB, Moharir K (2015) GIS-based quantitative morphometric analysis and its consequences: a case study from Shanur River Basin, Maharashtra India. Appl Water Sci 7(2)

Pande CB, Khadri SFR, Moharir KN, Patode RS (2017) Assessment of groundwater potential zonation of Mahesh River basin Akola and Buldhana districts, Maharashtra, India using remote sensing and GIS techniques. Sustain Water Resour Manag. https://doi.org/10.1007/s40899-017-0193-5

Pande CB, Moharir KN, Khadri SFR, Patil S (2018b) Study of land use classification in the arid region using multispectral satellite images. Appl Water Sci 8(5):1–11, ISSN 2190-5487

Pande CB, Moharir KN, Pande R (2018a), Assessment of morphometric and hypsometric study for watershed development using spatial technology—a case study of Wardha river basin in the Maharashtra, India. Int J River Basin Manag. https://doi.org/10.1080/15715124.2018.1505737

Pande CB, Moharir K (2018c), Spatial analysis of groundwater quality mapping in hard rock area in the Akola and Buldhana districts of Maharashtra, India. Appl Water Sci 8(4):1–17. ISSN 2190-5487

Pande CB, Moharir KN, Singh SK, Dzwairo B (2019a) Groundwater evaluation for drinking purposes using statistical index: study of Akola and Buldhana districts of Maharashtra, India. Environ Dev Sustain (A Multidisciplinary Approach to the Theory and Practice of Sustainable Development). https://doi.org/10.1007/s10668-019-00531-0

Pande CB, Moharir KN, Singh SK, Varade AM (2019b) An integrated approach to delineate the groundwater potential zones in Devdari watershed area of Akola district, Maharashtra, Central India. Environ Dev Sustain

Patode RS, Pande CB, Nagdeve MB, Moharir KN, Wankhade RM (2017) Planning of conservation measures for watershed management and development by using geospatial technology—a case study of Patur watershed in Akola District of Maharashtra. Curr World Environ 12(3)

Patode RS, Nagdeve MB, Pande CB (2016) Groundwater level monitoring of Kajaleshwar-Warkhed watershed, Tq. Barshitakli, Dist. Akola, India through GIS Approach. Adv Life Sci 5(24)

Reddy, Mahender D, Patode RS, Nagdeve MB, Satpute GU, Pande CB (2017) Land use mapping of the Warkhed micro-watershed with geo-spatial technology. Contemp Res India 7(3). (ISSN 2231–2137)

Printed in the United States
By Bookmasters